浙江省普通高校"十三五"新形态教材

中小企业对外贸易谈判

Foreign Trade Negotiation of Small-Medium-Sized Enterprises

主　编　姜继红

副主编　周美珍　金　晶

西安电子科技大学出版社

内 容 简 介

　　本书是一本围绕中小企业线上与线下对外贸易谈判活动,集文本、音频和视频于一体的新形态信息立体化教材。全书共分两大模块:传统国际贸易和跨境电子贸易,每个模块又分设具体任务。其中传统国际贸易下分九个任务:商务调查、商务接待、产品与服务、价格谈判、运输谈判、保险谈判、支付谈判、合同谈判和投诉谈判;跨境电子贸易下分六个任务:网店介绍、产品描述、售中服务、物流配送、售后服务和争端处理。每个任务从基础知识回顾、任务描述、典型对话案例、角色模拟练习、典型表达方式、外贸函电写作及翻译、外贸单据、学习效果评估等方面将线上与线下对外贸易谈判的基本技能技巧融入其中。

　　本书适合作为高职高专学校、成人培训学校和应用型本科学校商务英语、国际经济与贸易及跨境电子商务等专业的教材,也可作为对外贸易相关从业人员的自学教材。

图书在版编目(CIP)数据

中小企业对外贸易谈判 = Foreign Trade Negotiation of Small-Medium-Sized Enterprises /
姜继红主编. — 西安:西安电子科技大学出版社,2022.5(2024.7 重印)
ISBN 978 - 7 - 5606 - 6442 - 2

Ⅰ. ①中…　Ⅱ. ①姜…　Ⅲ. ①中小企业—对外贸易—贸易谈判—教材
Ⅳ.①F740.41

中国版本图书馆 CIP 数据核字(2022)第 060381 号

策　　划　　刘小莉
责任编辑　　刘小莉
出版发行　　西安电子科技大学出版社(西安市太白南路 2 号)
电　　话　　(029)88202421　88201467　　　邮　　编　　710071
网　　址　　www.xduph.com　　　　　　电子邮箱　　xdupfxb001@163.com
经　　销　　新华书店
印刷单位　　咸阳华盛印务有限责任公司
版　　次　　2022 年 5 月第 1 版　　2024 年 7 月第 2 次印刷
开　　本　　787 毫米×1092 毫米　　1/16　　印　张　17
字　　数　　400 千字
定　　价　　45.00 元
ISBN 978 - 7 - 5606 - 6442 - 2
XDUP 6744001-2
***** 如有印装问题可调换 *****

前　言

本书是浙江省普通高校"十三五"新形态教材第二批高职高专立项项目"中小企业对外贸易谈判"的成果教材。本书作者以"理实结合、工学一体、理论够用、实践为主"为编写原则，以精益求精的态度和深耕细作的实干精神，深入企业积极调研，细细打磨教材每一细节。本书的主要学习对象为高职高专涉外经贸类专业学生、商务英语专业学生、电子商务专业学生、跨境电子商务专业学生以及有志于从事线上线下国际贸易的社会人员。本书内容完整，覆盖范围广，体例新颖，系统专业，难易适中，深入浅出，实用性和可操作性强。

本书具有以下几个方面的特色：

(1) 本书采用"大案例，一案到底"的组织方式，即采用一个完整的项目"杭州红粉服饰有限公司关于服装国际贸易的谈判"贯穿全书。本书共设十五个具体任务，通过学生扮演对外贸易业务员展开线下商务调查、商务接待、产品服务及价格、运输、支付、合同和投诉等环节的谈判任务，并利用互联网开展跨境电子商务平台的网店介绍、产品描述、售中服务、物流配送、售后服务、争端处理等环节的任务。

(2) 知识点的组织采用"模块构建项目，项目构建任务，任务构建知识、技能、素质"的方式进行，使学生能够系统地掌握中小企业对外贸易谈判的知识，熟悉中小企业对外贸易谈判的整个过程，同时能够开展中小企业外贸业务中的各项谈判工作，丰富学生的谈判策略和技巧，切实提高学生对外贸谈判的实际操作能力。

(3) 本书配有音频、视频和 PPT 课件，并以二维码的形式嵌入书中，用户可通过手机或移动设备随扫随用，实现线上与线下学习资源的无缝衔接，既方便教师教学又方便学生自主学习。

(4) 本书增加了自我学习效果评估部分，方便教师与学生评价学习效果。

本书由浙江长征职业技术学院姜继红副教授担任主编，华南农业大学外国语学院英语系周美珍和浙江长征职业技术学院金晶担任副主编。具体编写分工如下：姜继红编写模块一中的任务 1～9，模块二中的任务 3～6 及附录，并负责书稿的整体设计与规划及后期的审稿工作；周美珍编写模块二中的任务 2，同时还负责书稿中、英文的审稿工作；金晶编写模块二中的任务 1。

本书是产学深度融合的成果，该书的编写得到了校企合作单位的大力支持。感谢纵森(香港)纺织品有限公司毛武兵总经理和东莞市智恒针织制衣有限公司为本书的顺利出版提供的大量国际贸易中 B2B 传统线下贸易的一手资料。感谢台州鑫明电子商务有限公司王超越经理和王益明经理、杭州莎莎跨境电子商务有限公司傅阳经理的大力支持，他们为本书的顺利出版提供了跨境电子商务线上贸易的一手资料，保证了本书的真实性与实用性。

限于编者水平，书中难免有疏漏之处，请专家和读者批评指正。

编　者

2022 年 1 月 20 日于杭州

Contents

Introduction to Business Negotiation

1. What is business negotiation?

Business negotiation is a method by which people settle differences. It is a process by which compromise or agreement is reached while avoiding argument and dispute.

In any disagreement, individuals understandably aim to achieve the best possible outcome for their position (or perhaps an organization they represent). However, the principles of fairness, seeking mutual benefit and maintaining a relationship are the keys to a successful outcome.

2. Stages of business negotiation

In order to achieve a desirable outcome, it may be useful to follow a structured approach to negotiation. For example, in a work situation a meeting may need to be arranged in which all parties involved can come together.

The process of business negotiation includes the following stages:

(1) Preparation

(2) Discussion

(3) Clarification of goals

(4) Negotiate towards a win-win outcome

(5) Agreement

(6) Implementation of a course of action

3. What can affect the ultimate outcome of business negotiation?

In any negotiation, the following three elements are important and likely to affect the ultimate outcome of the business negotiation: attitudes, knowledge and interpersonal skills.

(1) Attitudes

All negotiation is strongly influenced by underlying attitudes to the process itself, for example, attitudes to the issues and personalities involved in the particular case or attitudes linked to personal needs for recognition.

(2) Knowledge

The more knowledge you possess of the issues in question, the greater your participation in the process of negotiation. In other words, good preparation is essential. Do your homework and gather as much information about the issues as you can. Furthermore, the way issues are negotiated must be understood as negotiating will require different methods in different situations.

(3) Interpersonal skills

Good interpersonal skills are essential for the effective negotiations, both in formal situations and in less formal or one-to-one negotiations.

These skills include:

✓   Effective verbal communication

Reducing misunderstandings is a key part of effective negotiation.

✓   Listening

✓   Rapport building

Building stronger working relationships based on mutual respect.

✓   Problem solving

✓   Decision making

Learn more techniques to help you make better decisions.

✓   Assertiveness

Assertiveness is an essential skill for successful negotiation.

✓   Dealing with difficult situations

4. Top ten effective negotiation skills

A successful negotiation requires the two parties to come together and hammer out an agreement that is acceptable to both.

(1) Problem analysis to identify interests and goals

Effective negotiators must have the skills to analyze a problem to determine the interests of each party in the negotiation. A detailed problem analysis identifies the issue, the interested parties and the outcome goals.

(2) Preparation before a meeting

Before entering a bargaining meeting, the skilled negotiator prepares for the meeting. Preparation includes determining goals, areas for trade and alternatives to the stated goals. In addition, negotiators study the history of the relationship between the two parties and past negotiations to find areas of agreement and common goals. Past precedents and outcomes can set the tone for current negotiations.

(3) Active listening skills

Negotiators have the skills to listen actively to the other party during the debate. Active listening involves the ability to read body language as well as verbal communication. It is important to listen to the other party to find areas for compromise during the meeting. Instead of spending the bulk of the time in negotiation expounding the virtues of his viewpoint, the skilled negotiator will spend more time listening to the other party.

(4) Keep emotions in check

It is vital that a negotiator have the ability to keep his emotions in check during the negotiation. While a negotiation on contentious issues can be frustrating, allowing emotions to take control during the meeting can lead to unfavorable results.

(5) Clear and effective communication

Negotiators must have the ability to communicate clearly and effectively to the other side during the negotiation. Misunderstandings can occur if the negotiator does not state his case clearly. During a bargaining meeting, an effective negotiator must have the skills to state his desired outcome as well as his reasoning.

(6) Collaboration and teamwork

Negotiation is not necessarily a one side against another arrangement. Effective negotiators must have the skills to work together as a team and foster a collaborative atmosphere during negotiations. Those involved in a negotiation on both sides of the issue must work together to reach an agreeable solution.

(7) Problem solving skills

Individuals with negotiation skills have the ability to seek a variety of solutions to problems. Instead of focusing on his ultimate goal for the negotiation, the individual with skills can focus on solving the problem, which may be a breakdown in communication, to benefit both sides of the issue.

(8) Decision making ability

Leaders with negotiation skills have the ability to act decisively during a negotiation. It may be necessary during a bargaining arrangement to agree to a compromise quickly to end a stalemate.

(9) Maintaining good relationships

Effective negotiators have the interpersonal skills to maintain a good working relationship with those involved in the negotiation. Negotiators with patience and the ability to persuade others without using manipulation can maintain a positive atmosphere during a difficult negotiation.

(10) Ethics and reliability

Ethical standards and reliability in an effective negotiator promote a trusting environment for negotiations. Both sides in a negotiation must trust that the other party will follow through on promises and agreements. A negotiator must have the skills to execute on his promises after bargaining ends.

5. When do negotiations not work?

Even the best negotiators have difficulty at some point or another to make things work. After all, the process requires some give and take. Perhaps one party just won't budge and doesn't want to give in at all. There could be other issues that stall the negotiation process, including a lack of communication, some sense of fear, or even a lack of trust between parties. These obstacles can lead to frustration and, in some cases, anger. The negotiations may turn sour and ultimately lead parties to argue with one another.

When this happens, the best—and sometimes only—thing the parties can do is to walk away. Taking yourself out of the equation gives everyone involved a chance to regroup, and it may help both of you come back to the bargaining table with a cool and fresh mind.

Traditional International Trade

Task 1　Business Investigation

知识目标

◆ 学习国际贸易中商务调查的内容
◆ 学习国际贸易中商务调查的渠道
◆ 学习相关商务沟通的口头英文表达方式
◆ 学习相关商务沟通的书面英文表述技巧

能力目标

◆ 能够准备商务调查材料
◆ 能够顺利开展商务调查
◆ 能够使用双语顺畅地进行商务口头沟通
◆ 能够处理商务调查往来信函

Basic Knowledge Review

1. What does business investigation involve in the international trade?

Business investigation is a thorough investigation done by businesses to reveal any potential risks or fraud. Business investigation comprises both preventive and countermeasure action. By conducting investigation, a business can prevent risks from happening and minimize loss if a fraudulent action ever takes place.

When considering a new market or buyer/partner, take time to conduct due diligence. It could help you avoid costly obstacles. Investigate the political, economic, and financial conditions of the market. Carefully select partners and buyers to ensure a successful and profitable relationship.

Good due diligence will help protect your company from problems, loss, and liability. Generally speaking, there are two factors which evaluate potential buyers or partners carefully.

(1) Country risk

If you are interested in a new market, look closely at the political, economic and business environment to ensure it is a good trade for your company. Country factors to research include:

✓　Political stability
✓　Foreign exchange risk
✓　Economic stability

✓ Legal system
✓ Intellectual property protection laws
✓ Banking structure
✓ Tax implications

(2) Buyer risk

Before concluding an export deal or agreement, conduct due diligence on the foreign entity to make sure doing business has the best chance of success for your company and is legal.

A bad choice can result in lost market opportunities, financial loss, legal and liability issues, or a damaged company/product reputation abroad.

Vetting will help ensure the buyer is legitimate and credit-worthy. Where can you get the information to properly evaluate a potential buyer or foreign partner?

✓ Local international commercial service

If you need a background check on a foreign company, consider buyers' local international commercial service.

✓ Banks

Bankers have access to vast amounts of information on foreign companies and are usually very willing to assist corporate customers.

✓ Commercial firms

Commercial firms can provide credit information on foreign firms. This information may also be in directories of international companies.

✓ Business libraries

Several private-sector publications list and qualify international companies. There are also many directories devoted to specific regions and countries.

✓ Foreign embassies

The commercial (business) sections of most foreign embassies have directories of companies located in their countries.

✓ Internet

You can often find company information through your favorite search engine. However, pay attention to the sources of the information you find, because search engine results may not always be recent or accurate.

2. How do we achieve the goal of business investigation successfully?
While no two corporate investigations are the same, there are some standard steps that good investigation firms will undertake no matter what the challenge is.

(1) Debrief and evaluate
We meet with the client for full debrief of the situation. We evaluate the scope and extent of the possible fraud or corruption. Does it merit a full investigation, or is there a way to get to the answers quickly and easily? Is there enough basic evidence to proceed?

(2) Background checking
We check the backgrounds of the individuals and/or companies involved, looking for evidence of fraud or corruption. This may involve employees, management or outside contractors.

(3) Make a plan and get prepared
The key to successful investigations—and evidence that holds up in court—is mapping out a plan and making sure you have all your documentation, materials, information and contingency plans in place before you get started.

(4) External investigation
It often makes sense to start investigating witnesses and organizations outside the company being investigated, so that when the investigation becomes internal, you have more information at your disposal.

(5) Document specific fraudulent behavior
It's not enough to have "feeling" or circumstantial evidence. Good investigators gather detailed, specific information proving fraudulent behavior on the part of an individual or organization.

(6) Communicate with the client
Of course client communication is critical at all steps of the investigation, but at this point it's important to review information, lead possible conclusions and determine next steps.

(7) Refine investigation
Based on information gathered, meetings with the client, and investigation objectives, refine initiatives and activities which adapt to the overall goals.

(8) Prepare the report
Documenting the actions taken, information gathered, and recommendations for specific next

steps is the best way to make your investigation useful and effective for the client.

Nowadays, the internet has been the primary medium of communication for people of this modern generation. As a part of this development, professional e-mails are utilized to introduce themselves and expand their connections. Since you are going to introduce yourself, it is just rational to say that you are a stranger to your recipient's eyes. Similarly, to clearly discuss the fundamentals of introduction e-mails, it would be great to show you how it looks like first without any further ado.

3. What is an introduction e-mail?

When you want to introduce yourself, your team, or your company to someone through e-mails, you send an introduction e-mail. An introduction e-mail etiquette, as the name suggests, is sent primarily to introduce someone or something usually to the recipient of such e-mails.

The contents of an introduction e-mail vary from one person to another. Typically, an introduction business e-mail contains the name of the person (persons) or organization being introduced, the affiliations of the ones being introduced, and the reason or reasons for the introduction. However, as mentioned, it may completely vary from one e-mail to another.

If a person introduces the business to a certain recipient, he/she will include the contents mentioned above, and may also include other details such as the service or products being provided by such business.

Five tips for making your business introduction letters stand out!

(1) Do your research. Research is absolutely essential to your letter, and tells your prospective employer, customer, or partners that you know what you're talking about. Explain what special skills or unique selling proposition you can offer your recipients. Explain how your services or skills will enhance the incredible job they're doing and blow their competition out of the water.

(2) Understand their company culture and brand. If the business you are approaching doesn't use formal language, then you'd better write the letter using terms and words that they associate with their company. You can get a feel for how a business views itself from the carefully chosen language they use on their website, in advertisements, and even press releases.

(3) Keep it short and sweet. You want the business to contact you for more information on how you can enhance what they do. Aim for powerful, impactful language, but keep a business introduction letter under 400 words.

(4) State your purpose before anything else. Let the reader know you are writing to introduce

yourself and your skills/talent/products to their company, and why your skills/talent/products are a natural fit.

(5) Always close with a call to action. Let your readers know exactly what you're hoping they'll do, such as give you a phone call, schedule an appointment, or join your mailing list. Let them know you'll also be following up on a specific date. When waiting for a follow-up call or e-mail, be patient and give them at least three business days, preferably at least five, to get back to you. You are not expected to appear impatient when you're the one who will benefit most when they choose to get in touch with you.

Task Descriptions

Business Investigation

　　杭州红粉服饰有限公司是一家集专业设计、研发、加工、生产和贸易为一体的现代化服装企业，经营各类服装服饰，多年来秉承"质量第一，信誉至上"的经营理念，一直服务于亚洲、欧美等地区客户。美国斯特朗服装有限公司是杭州红粉服饰有限公司在广州的中国进出口商品交易会上结识的新客户。Dicky 是杭州红粉服饰有限公司业务员，主要负责美国市场的开发与维护。Betty 是美国斯特朗服装有限公司采购部经理。

　　接下来的对话就是围绕杭州红粉服饰有限公司与美国斯特朗服装有限公司一系列贸易活动展开的。

Typical Dialogues

◇ **Dialogue Ⅰ**

　　这段对话的内容是：杭州红粉服饰有限公司外贸业务员 Dicky 向美国斯特朗服装有限公司采购部经理 Betty 了解美国服装市场需求，为建立进一步的商业关系做准备。

A：杭州红粉服饰有限公司外贸业务员 Dicky

B：美国斯特朗服装有限公司采购部经理 Betty

A：I just wonder if Maxi dress might be our premier collection this year.

A：我在想沙滩裙是否可以成为我们今年的主打款。

B：Sure it will be. The popularity of Maxi dress is still on the rise according to the survey result from our US market, and it has become almost a necessity and standard in virtually every lady's wardrobe and closet.

B：当然可以。根据我们对美国市场的调研，沙滩裙的流行势头一直不减，它已经成为几乎所有女性衣橱中的必备单品。

A：Well, that's pretty good. If we can make some improvements on the previous design, it can be the money maker.

A：那真是太好了。如果我们可以在先前的设计上做一些亮点，这款裙子就可以成为我们今年的摇钱树了。

B：Yes. In addition to the fashion design, price is also an important variable to build our customer base.

B：是啊。除了时尚的设计，价格也是构建客户基础的重要因素。

A：That's true. So we shall hold a further discussion on setting the price after the design has been worked out.

A：没错。所以我们要在设计出来后再深入讨论一下定价。

B：OK.

B：好的。

■ **Notes:**

1. Maxi dress：沙滩裙

2. premier collection：主打款

3. variable：变量

◇ **Dialogue Ⅱ**

这段对话发生在杭州红粉服饰有限公司外贸业务员 Dicky 与美国斯特朗服装有限公司采购部经理 Betty 之间。双方讨论产品价格对市场的影响。

A：美国斯特朗服装有限公司采购部经理 Betty

B：杭州红粉服饰有限公司外贸业务员 Dicky

A：I went over the sales figures in the market research. It seems there's only one conclusion.

A：我看过市场调查报告中的销售数字了。看起来结论似乎只有一个。

B：What's that?

B：是什么呢？

A：It seems this market is sensitive to price.

A：市场似乎对价格很敏感。

B：So, shall we start at a low introductory price?

B：那我们开始时是不是要定一个比较低的市场推广价呢？

A：I think that would be a good idea. Price is an important variable in our market, so we can use it to build our customer base. We can bring the price up after our customer base is consolidated.

A：我想这个主意不错。价格在我们的市场上是一个重要的变量，所以我们可以利用价格建立我们的顾客基础。我们可以在顾客基础稳固之后再把价格提高。

◇ **Dialogue Ⅲ**

这段对话的内容是：杭州红粉服饰有限公司外贸业务员 Dicky 向美国斯特朗服装有限公司采购部经理 Betty 了解美国本土文化习俗。

A：杭州红粉服饰有限公司外贸业务员 Dicky

B：美国斯特朗服装有限公司采购部经理 Betty

A：I thought tailored suits are well received in the US, but actually it's not so popular as that in Europe.

A：我原以为定制西服在美国会很热卖，但事实证明它并没有像在欧洲那么畅销。

B：That's for sure. Americans are far more casual in dressing.

B：那是当然了，美国人更喜欢随意的穿搭。

A：It seems they care more about comfort.

A：美国人看起来更讲究的是舒服。

B：Yes, even on Wall Street, men are often seen wearing suits and big sneakers on their feet.

B：是啊，即便是在华尔街，常常都会看见男士们上身穿西装，脚上却踩着一双大球鞋。

A：I once saw an American business woman wearing designer suit, but with a pair of sneakers.

A：我曾遇到过一个美国职业女性穿着名牌套装却搭配了一双球鞋。

B：Actually, it's not so surprising to see Americans dressing like that.

B：事实上美国人有这样的穿搭是很稀松平常的。

A：I finally understand why sneakers sell so well in the US.

A：我终于明白了为什么运动球鞋在美国会卖得那么好。

■ **Notes：**

tailored suit：考究的套装

◇ **Dialogue Ⅳ**

这段对话发生在杭州红粉服饰有限公司外贸业务员 Dicky 与美国斯特朗服装有限公司采购部经理 Betty 之间。双方讨论了在商务活动中跨文化交际的相关问题。

A：美国斯特朗服装有限公司采购部经理 Betty

B：杭州红粉服饰有限公司外贸业务员 Dicky

A：I think there is some relationship between business and culture. What do you think?

A：我想做生意和文化是有关系的，你看呢？

B：Yes, there is indeed. I think cultural differences occur in business on three levels.

B：是的，确实有关系。我觉得文化差异在商业上可表现为三个层面。

A：Three levels? Go ahead.

A：三个层面？愿听其详。

B：The first level is etiquette; the second level is behaviors and actions; and the third is the level of core morals, beliefs and values.

B：第一个层面表现在礼节上；第二个层面表现在行为举止上；第三个层面则表现在核心道德、信仰以及价值观上面。

A：Great, deep-going analysis.

A：好，分析得很深刻。

B: Some people often make the mistake of seeing foreign cultures through their own habitual cultural standard.

B: 有些人习惯于以自己的文化标准来看待外国文化，因而经常犯错误。

A: So, you once experienced this misunderstanding?

A: 这么说，您曾经经历过这样的误解？

B: Yes. In my professional career, there are some cases in which cultural differences have led to misunderstandings in business.

B: 是的。在我的职业生涯中，就有过因文化差异而导致生意上出现误解的情况。

◇ **Dialogue V**

这段对话的内容是：美国斯特朗服装有限公司采购部经理 Betty 向杭州红粉服饰有限公司外贸业务员 Dicky 了解公司的基本情况。

A: 美国斯特朗服装有限公司采购部经理 Betty

B: 杭州红粉服饰有限公司外贸业务员 Dicky

A: We are deeply impressed by your summer dress wear, and just wanna know more about your company. How long have you been in this line of business?

A: 你们的夏季裙装很不错，所以我们想更多地了解一下贵公司。你们在这个行业发展有多久了呀？

B: Hangzhou Pinklife Fashion co. Ltd. was established in 2000, and has now become a medium-sized company with major clients in Europe. We are planning to expand our market to north America.

B: 杭州红粉服饰有限公司成立于 2000 年，现已发展成为一家中型企业，客户群主要在欧洲。我们正打算把市场拓展到北美市场。

A: Would you mind telling us your regular bank?

A: 方便告知一下你们的往来银行吗？

B: Of course. Our reference bank is Bank of China, Hangzhou Branch.

B: 当然可以，我们的参考银行是中国银行杭州分行。

A: Thanks a lot. You can also obtain information about us from Bank of America.

A: 非常感谢。您也可以从美国银行了解到更多关于我公司的信息。

B: All right. Hope everything goes well.

B: 好的。愿一切顺利。

A: Surely it will be.

A: 一定会的。

■ **Notes：**

regular bank：往来银行

reference bank：参考银行

两者均指平时业务往来较多、可提供相关资料的银行。

Roles Simulation

Suppose you are one of them in the following conversation. Try to read aloud and practice the underlined sentences.

这段对话的内容是：杭州红粉服饰有限公司外贸业务员 Dicky 向中国银行杭州分行的工作人员调查美国斯特朗服装有限公司的资信状况。

A：中国银行杭州分行的工作人员

B：杭州红粉服饰有限公司外贸业务员 Dicky

A：Good morning, <u>this is Consultant Department of Bank of China, Hangzhou Branch.</u> What can I do for you?

A：早上好，<u>这里是中行杭州分行咨询部。</u>有事吗？

B：Well, <u>we have got a new partner and planned to sign a contract at the end of this month. I'd like to know more about that company.</u>

B：<u>我们有个新的贸易伙伴，计划在月底签约，因此我想更多地了解一下这家公司。</u>

A：You mean you want us to make a status inquiry?

A：您的意思是让我们为您做个资信调查？

B：Yes, <u>I want to get the credit information about our cooperative partner.</u>

B：是的，<u>我想知道合作伙伴的资信情况。</u>

A：OK. I'll see what I can do for you.

A：好吧，我来看看能为您做些什么。

B：<u>Will making a credit investigation be difficult?</u>

B：<u>做资信调查很难吗？</u>

A：No. It's not complicated. The only thing for you to do is fill out a form and sign it.

A：不，这并不复杂。您唯一要做的就是填张表并签上自己的名字。

B：And then?

B：然后呢？

A：<u>Then we will send it to our branch office or agent in America by the fastest way. They will do as your request, then send the report back. Of course, it is done in a confidential way, and we hope you will keep it confidential.</u>

A：<u>然后我们将以最快的速度将表格传给美国办事处，他们将依据您的要求办理，然后将报告返还。当然这是在保密状态下进行的，还希望你们可以保密。</u>

B：Yes, of course we will. What about the service fee?

B：我们当然会保密的。那么要付多少手续费呢？

A：For our long-term clients we charge only direct fees like telephone or fax fee.

A：对长期客户，我们只收取电话、传真费用。

B：Thank you.

B：谢谢！

(A week later)

(一个星期以后)

B：Hello. Do you have the status report?

B：您好，资信报告出来了吗？

A：Yes. <u>We have just completed our inquiries concerning the firm mentioned in your trust form.</u>

A：有。<u>我们刚刚完成你们委托表中所提到公司的资信调查。</u>

B：Any troubles?

B：有什么麻烦吗？

A：No, this is a favorable reply. It's a private firm and enjoys good reputation in its area. <u>As the credit report shows, it's always punctual to meet its commitments. It seems to be safe to do business with them.</u>

A：没有，结果很有利。这是一家私人企业，在该地区享有良好的声誉。<u>资信报告显示，他们总是能及时履行义务。看来和他们做生意应该是安全的。</u>

B：Great. <u>Now we can negotiate with them with assurance.</u> Thank you for your information.

B：太好了！<u>现在我们可以放心地和他们洽谈了。</u>　谢谢您提供的信息。

❤❤ Typical Expressions

1. How is the market situation in your country? 贵国的市场状况如何？

2. People's Bank of China advises that the company was originally established in 2001 by Claliris Smith, which deals with the export and import of pharmaceutical industry. 中国人民银行通知，该公司于 2001 年由克拉里斯·史密斯始创，主要从事医药行业的进出口业务。

3. We are informed that the company is a newly-formed corporation founded in September of 2018 with an authorized capital of $50,000 but with the actual amount of $500 flown in. 被调查公司的基本情况如下，2018 年 9 月创立，注册资本 50,000 美元，股东实际缴纳金额 500 美元。

4. We have just completed our inquiries concerning the firm mentioned in your trust form. 我们刚刚完成你们委托表中所提到公司的资信调查。

5. We regret our inability to tell you anything positive concerning the firm mentioned in your letter of the 30th ult. 很抱歉，对于上个月 30 号来函要求调查的公司，我们无法提供其确切信息。

6. As the credit report shows, it's always punctual to meet its commitments. 资信报告显示，他们总是能及时履行义务。

7. The firm under inquiry enjoys a high reputation in the business circles for their punctuality in fulfilling obligations. 该被调查公司在商界享有很高的声望，他们总能很及时地履行其义务。

8. Of course it is done in a confidential way, and we hope you will keep it confidential. 当然这是在保密状态下进行的，还希望你们可以保密。

9. As far as we know, they are sound enough, but we have no certain knowledge of their true financial position. 据我们所知，他们非常可靠，但对其真实的财务情况我们不甚了解。

10. They are rated as A1 company. You can deal freely with little risk. 他们被定为 A1 级公司。你们可放心地与之交易，没有什么风险。

Foreign Trade Letters

◇ Sample I   Introduction e-mail

The following is a letter introducing Martin's business to Thomas's business partner by Gmail.

M Gmail

Introduction LFC   2021 年 5 月 31 日 上午 8:59
XXXXXX
收件人：XXXXXX
抄送：XXXXXX
Good morning Chris，

I hope all is well with you and you are enjoying the UK summer months.

I would like to establish direct communications between you and our factory JV partner Mr Martin Mao.

Damon is still involved in this business through Design and Technical Departments but I stopped my daily involvement in operational aspects and focus on our own Brand Development Business.

I think I might have told you, but Martin operates our increased capacity woven factory and I believe his new pricing structure is something that will fit with LFC.

Martin will take on communications from here on and keep me copied in, but when next you are in China it would be a pleasure to meet up.

All the best and regards from Damon.

XXXXXX
XXXXXX HK Ltd
XXXXXX China Ltd
w: www.xxxxxx.com
t: +86 XXXXXXX (China)

---

t: +49 XXXXXXX(Germany)

Think 'Green' before you decide to print this mail and attachments. Recycle your waste at home and at work!

Disclaimer: The information in this e-mail is confidential and is legally privileged. It is intended solely for the addresses. If this e-mail is not intended for you, you cannot copy, distribute or disclose the included information to anyone and you are requested to delete this mail. While all reasonable steps have been taken to ensure the accuracy and integrity of all data transmitted electronically, Pattern and Stitch Hong Kong Ltd and Pattern and Stitch China Ltd, do not accept liability if the data, for whatever reason, is corrupt or does not reach its intended destination.

Please note that all comments and opinions stated by the author of e-mails sent to designated recipients: are "WITHOUT PREJUDICE" and are "PRIVATE AND STRICTLY CONFIDENTIAL".

■ **Notes：**

　　Disclaimer(免责声明)：是 Gmail 邮件的系统自动发送的。Gmail 是 Google 的免费网络邮件服务，提供 15G 存储空间。截至 2015 年 3 月，Gmail 界面支持 72 种语言：Arabic, Basque, Bulgarian, Catalan, Chinese (simplified), Chinese (traditional), Croatian, Czech, Danish, Dutch, English (UK), English (US), Estonian, Finnish, French, German, Greek, Gujarati, Hebrew, Hindi, Hungarian, Icelandic, Indonesian, Italian, Japanese, Kannada, Korean, Latvian, Lithuanian, Malay, Malayalam, Marathi, Norwegian (Bokmål), Odia, Polish, Punjabi, Portuguese (Brazil), Portuguese (Portugal), Romanian, Russian, Serbian, Sinhala, Slovak, Slovenian, Spanish, Swedish, Tagalog (Filipino), Tamil, Telugu, Thai, Turkish, Ukrainian, Urdu, Vietnamese, Welsh and Zulu etc. Gmail 界面设有 100 多种虚拟键盘、语言自助转换和输入法(其中还包括手写输入)，方便不同语种的用户使用。据统计，截至 2019 年 10 月，全世界共有 15 亿 Gmail 用户。

◇ **Sample Ⅱ　Reply letters from buyer Sunny**

-----Forwarded message----

发件人：XXXXXX

Date：2021 年 9 月 6 日周三 下午 9：30

Subject：Re：PF-STYLE#PF1703K-GRANGER TOP, STEINMART**NEW INQUIRY**

To: XXXXXX

Cc: XXXXXX

Hi Limy,

This is Sunny, thank you very much for your e-mail. I am taking of Amelotto ITL Co., wovern, cut and sewing business.

We got samples and got reference about your products.

We have most of customer need approval factory by BV or others before they order. For that reason, I should visit your factory and know your team, my schedule time to visit your factory will be between Oct.20-25.

I am looking forward to meeting your team and your boss Martin in China, hope we will feel confidence to work with you soon.

Before my visiting, we will keep touch with you if there are enquirers.

Best Regards
Sunny

ABC Oral Practice

Forming a group of two or three partners. Try to work out a dialogue related to the theme of this unit and then perform it in class.

Self-Assessment

Evaluate your practice by marking the following each corresponding item and work out the total scores.

考核情境(分值)	考核要求(分值)	得　分
语音语调(20分)	发音准确，声音清晰(10分)	
	语调自然，语速自然流畅(10分)	
内容(30分)	内容完整(10分)	
	内容表述符合给定情境(10分)	
	前后内容逻辑一致，内容之间无矛盾(10分)	
语言描述(30分)	用词规范、准确，无语法错误(20分)	
	表述无歧义(10分)	
其他(20分)	无雷同状况(10分)	
	无严重偏题(10分)	
合　计(100分)		

Task 2   Business Reception

知识目标

◆ 学习国际贸易中商务接待的过程
◆ 学习国际贸易中商务接待的文化
◆ 学习相关商务沟通的口头英文表达方式
◆ 学习相关商务沟通的书面英文表达技巧

能力目标

◆ 能够提前做好商务接待准备工作
◆ 能够顺利安排商务接待各个环节
◆ 能够使用双语顺畅地进行商务口头沟通
◆ 能够处理商务接待往来信函

Basic Knowledge Review

1. Tips to dress for business meeting

Dress appropriately for business. It is very important that you wear comfortable shoes. Check weather conditions in advance to determine type of wardrobe to include.

First impressions are very important. Each person will be ambassadors for their respective community and business; therefore, it is important that individuals conduct themselves appropriately at all times. Remember, you cannot take back your first impression.

Learn how to dress for success for business meetings with these tips.

(1) Appropriate attire

Employees and managers both agree that appropriate business attire makes people more lucrative. The professionals must always dress up in accordance to their professionalism. A suit is the best business meeting attire.

Even though a T-shirt or jeans may be appropriate, suit has its own way to make you look highly productive.

(2) Business casuals

Business casuals usually is a combination of classic formals with trendy modern patterns and crop wear pieces.

Business casual for women can be Capri pants or regular well fitted pants or knee length pencil skirts. On it tailored shirts and blouses would be apt.

Never forget to add a blazer or even jacket which goes the same for men and women as well.

Business casual attire for men can be a combination of pants, sweater and a jacket which is another version of business casual yet leaves you looking professional.

(3) Extreme skin exposure is not allowed

Women flashing lot of skin during a meeting will be misinterpreted and it can distract people's attention.

On the other hand, men wearing shorts and T-shirts will be taken as immature, incompetent and unprofessional by others at the meeting.

(4) Perfect blouse for women

For women, a knit or silk blouse or a close weave sweater will be apt. These will help you for not being too revealing.

Ensure that the top you pick must flatter your upper body figure. Go for solid colors and rich hue pattern on silks would be ideal.

(5) About the tie

No doubt ties are pretty essential if men choose the classic formal style of apparel. It will make you look professional.

When you go for business casuals, ties are not necessary. Make sure you are teaming up all the right pieces for your final outfit.

Choose a good quality tie made of good quality fabric such as silk. Go for subdued patterns or color and not any character prints.

(6) Match the belt with suit or the shoes

Not any piece of your apparel should be missed out and mismatch with your entire success look, when choosing a belt go for a color that matches your suit or the color of your shoes.

(7) Match trousers to jacket

In business casual, if you are going for the two-piece suit, it is the best. If you are buying trousers separately, make sure it matches the color of your jacket.

(8) Go for the perfect fit

Whether you choose to wear a business casual or formal suit, the crucial thing to keep in mind is that all the pieces of the proper attire must fit you perfectly.

It looks really unsuitable when your clothes are tight for you, conversely, when your clothes are loose for you, it might appear as if you have borrowed them, or were the clothes before you lost your weight and you forgot to buy new ones.

(9) The right fit

Pay attention to the cuts and fit of the business formal attire and make sure it suits your body shape.

When you wear a suit, the shoulders must sit square and the trouser end should be right before the shoe line.

(10) Consider khakis

In business wear, khakis are common for both men and women. It is really ideal to stand out in a meeting. Make sure the khakis are well ironed and wrinkle free.

(11) Match the socks with your trousers

It goes for both the sexes. Black color socks are versatile and the most common ones, however, you may want to match your sock's color with the color of your shoes or trousers so that it blends with your entire outfit.

Avoid socks that are colored or have flashy patterns or even a white colored one.

(12) Hosiery for women

Sometimes it is necessary for women to opt for hosiery for skirts, or some trousers as well. You do not expose a lot of skin with the help of these and make others to pay attention to what you have to present and say other than anything else.

(13) Attention to details

Your clothes should be clean, fresh as well as ironed. Ensure you do not have any loose buttons or unwanted creases or even stains. Pay attention to grooming; nails must be clean and short.

For women, avoid wearing extremely funky nail paints and if you choose to apply nail paint then wear a lighter shade, suiting your outfit. Avoid chipped nails. Men should polish their leather shoes.

(14) Your shoes speak a lot about you

Women can wear the classic low-heeled black shoes. But for one to look and feel more confident, high-heeled shoes will be ideal.

It is not necessary your shoes have to be black; you can go for the color that will compliment and match the rest of your outfit.

Men must always go for leather oxford shoes. Make sure it matches the same color as the trouser or a shade darker.

(15) It's incomplete without a hairstyle

Your entire professional business attire will fail if you lack the perfect hairstyle that goes well with your apparel. People in the meeting should not be distracted by your scruffy, messy hair.

Men must trim the hair well and women must take care that their hair is not falling over their face, as it will just not distract you but the others in the meeting.

Women can select from tying their hair into a loose ponytail or a bun. These can be easily done by using a hair band and totally reflects elegance.

(16) The perfect amount of makeup

Make sure you are wearing a light perfume, appropriate for daytime and is not overpowering. The overpowering smell of a perfume may linger around the room for hours and disgust the people in the meeting. Women must go for natural look while doing their makeup for a meeting or work.

(17) Accessories to display elegance

One must wear a watch that is simple yet elegant to display success and professionalism. In case you have colorful watches, ensure that they match your casual business attire. But a timeless leather banded watch is ideal.

Women should avoid wearing earrings and necklaces that are huge or flashy. A single, simple and elegant piece of jewelry on one will be enough to speak taste. Avoid being a hipster at the meeting.

(18) Regard any social or cultural influences

If the business meeting is with international clients, you need to do a little research as to what is appropriate and what is not. It implies what attire their business dress culture considers professional; it can mean covering up a little more or upgrading your overall look for decorum.

(19) Consider your seniors or clients

Try to dress up to the same level as your seniors or your clients. For an instance, if the meeting is with high level executives who wear and prefer suits, dress up in the same manner.

(20) Stand by comfortable clothes

You would not want synthetic materials as they trigger sweat, which is unacceptable at the time of a business meeting. Opt for crease free cotton material clothes which will allow your skin to breathe.

(21) Proper handbag

You undoubtedly require a handbag as you do not want to enter a meeting room with all your personal stuff hanging out from your hand. To look organized is highly essential if you want to appear professional. Men and women must invest in a good corporate handbag which is timeless and stylish. Select a black color handbag or a color that matches your outfit for the meeting.

(22) Be confident and feel comfortable

Dressing up for a business meeting is about developing your own style statement. Since you are representing a position with specific power, you must not take risks with the business opportunities by wearing odd clothes. It doesn't confine you to only wearing suits that are sometimes uncomfortable or shirts that you dislike. Discover your best suit for a formal attire in which you feel comfortable and self-assured.

Following the above-mentioned business meeting attire tips, you will recognize the importance of being dressed for success.

For a business meeting, go for the suitable style for you, let business clothes speak for you and leave an impression on the others of your field of you as a credible and successful expert in the field. The next time you are preparing for your meeting, take some time to prepare to look your best at the meeting as well.

2. How should we treat our customer in the meeting room?

It is polite and hospitable to prepare fruits, food and drinks during meeting with customers in the meeting room. It is healthy and easy to eat fruits. Choose good coffee the western like such as Starbucks. As for good drinks, Cola and orange juice are ok.

✍ *Task Descriptions*

美国斯特朗服装有限公司采购部经理 Betty 受公司委托决定亲自到中国来考察。杭州红粉服饰有限公司委派外贸业务员 Dicky 作好此次美国客户考察的全程接待业务。

Business Reception

接下来的对话就是围绕 Betty 在杭州参观杭州红粉服饰有限公司的一系列活动展开的。

Typical Dialogues

◇ **Dialogue** Ⅰ

这段对话发生在杭州萧山机场，杭州红粉服饰有限公司外贸业务员 Dicky 在出站口接待前来考察的美国斯特朗服装有限公司采购部经理 Betty。

A：杭州红粉服饰有限公司外贸业务员 Dicky

B：美国斯特朗服装有限公司采购部经理 Betty

A：Excuse me, but are you Miss Betty from Stelang company?

A：不好意思，请问您是斯特朗公司的 Betty 女士吗？

B：Oh, yes, I am.

B：嗯，是我。

A：How do you do, Miss Betty? I'm Dicky from Hangzhou Pinklife Fashion co. Ltd. Welcome to Hangzhou, and I'm here to meet you.

A：您好，Betty 女士。我是杭州红粉服饰有限公司外贸业务员 Dicky。欢迎来杭州。今天由我来负责接待您。

B：How do you do, Dicky? It is so kind of you to pick me up.

B：你好，Dicky。感谢您来接机。

A：Thank you for coming all the way from U.S. to visit our company. How was the flight?

A：也感谢您不远万里从美国专程来访我公司。旅途还好吗？

B：It was a pleasant trip, but a little bit exhausting.

B：旅途很好，就是有一点累。

A：Well, do you have all the luggage here, and shall we go straight to the hotel so that you can take a rest to recover from the jetlag?

A：哦，您的行李都在这吗？那我们直接去酒店吧，这样您就可以尽快休息以便倒时差。

B：That would be great. You are so considerate and thoughtful.

B：太棒了。您考虑得真周到。

A：Let me help you with the case.

A：我来帮您搬行李吧。

B：No, thank you, I can handle it myself.

B：不，谢谢您，我自己可以拿。

A：All right. The car is outside at the exit, and this way please.

A：好吧。车子在外面出口处，请走这边。

B：OK.

B：好的。

◇ Dialogue Ⅱ

这段对话发生在去杭州凯悦酒店的路上，是杭州红粉服饰有限公司外贸业务员 Dicky 和美国斯特朗服装有限公司采购部经理 Betty 之间的谈话。

A：杭州红粉服饰有限公司外贸业务员 Dicky

B：美国斯特朗服装有限公司采购部经理 Betty

A：Miss Betty, is it your first to visit China?

A：Betty 女士，这是您第一次来中国吗？

B：No, I went to Shanghai with my family three years ago, and we'd been there for a week. Well, it is the first time to come here, and Hangzhou is really a very beautiful city.

B：不，三年前我们全家去上海旅行待了一周，但来杭州是第一次，杭州确实是一座很美的城市！

A：Yeah, it truly is. This time you are going to live in Hyatt Hotel which is located in the city center and is close to the Westlake. You can enjoy the beautiful scenery of the Broken Bridge, Baochu Pagoda and Leifeng Pagoda inside the hotel room. Besides, you can also wander around the Westlake after supper.

A：是，这次给您安排的是位于市中心西湖边上的凯悦大酒店，在房间里就能远眺断桥、保俶塔和雷峰塔等著名美景，晚上还能去湖边漫步。

B：Fantastic! I can enjoy the Westlake in person. Thank you for your kind arrangement. Is it far from here?

B：西湖，终于可以亲眼欣赏到了，好棒，真的很感谢你们的周到安排。这儿过去远吗？

A：You're welcome. It's not so far, and takes about an hour. But as it is the rush hour, there will be the traffic jam. The navigation shows that the current road condition is OK. It should be at most one and a half hours to get to the hotel. You can take a nap, and I will call you when we are there.

A：不客气。不远，一般 1 个小时就可以到了，但现在是晚高峰时间，估计路上会有点堵，导航显示目前的路况总体还行，那应该最多一个半小时可以到酒店。您可以在车上稍微睡一会儿，到了我叫您。

B：OK, thank you.

B：好的，谢谢。

◇ Dialogue Ⅲ

这段对话发生在杭州凯悦酒店，杭州红粉服饰有限公司外贸业务员 Dicky 与杭州凯悦酒店前台接待工作人员为美国斯特朗服装有限公司采购部经理 Betty 办理入住手续。

A：杭州凯悦酒店前台接待工作人员

B：杭州红粉服饰有限公司外贸业务员 Dicky

C：美国斯特朗服装有限公司采购部经理 Betty

A：Good afternoon! Welcome to Hyatt.

A：您好！欢迎光临杭州凯悦酒店。

B：Good afternoon! I've booked a double a week ago.

B：您好！一周前我订了个标间。

A：Sorry for a while. Let me have a check. May I have your name?

A：请稍等，我查一下，请问您的名字是？

B：Allen Betty.

B：艾伦·贝蒂

A：Oh, yes, here is the booking. Please fill in this registration form first. Do you have any special requirements for the room?

A：嗯，是的，这有您的预定。请先填一下这张表格。请问对住宿有特别的要求吗？

B：Can I have the lake view room?

B：请安排湖景房好吗？

A：Sure, here is your room key, and your room is 606 on the sixth floor. Please go this way to take the lift. Wish you a good stay here!

A：好的。您的房间在 6 层的 606 客房，这是您的门卡，请走这边坐电梯。祝您下榻愉快！

B：OK, thank you.

B：好的，谢谢。

(After entering the room)

(进入客房后)

C：Wow! It's really a nice room. Thank you so much, Dicky.

C：哇！这个房间真的不错，Dicky，谢谢你。

B：You're welcome, and I'm so happy you like it. You can take a rest for a while, and shall we meet at the dining hall on the third floor after an hour. Here is your seven-day schedule here in Hangzhou. Please have a look, let us know if it is OK.

B：很高兴您能满意。您先稍事休息，1 个小时后我们在 3 楼餐厅见。这份是您在杭州这 7 天的行程安排，你看看是否可以，如有不便我们可以再调整。

C：OK, thank you. See you later.

C：好的，谢谢。稍后见。

◇ Dialogue Ⅳ

这段对话发生在美国斯特朗服装有限公司采购部经理 Betty 即将离开杭州的前一天晚上，是杭州红粉服饰有限公司外贸业务员 Dicky 代表公司正式邀请 Betty 一起共进晚餐时的对话。

A：杭州红粉服饰有限公司外贸业务员 Dicky

B：美国斯特朗服装有限公司采购部经理 Betty

A：Well, here comes Dongpo Meat. Please help yourself, Betty.

A：Betty，这是东坡肉，请慢用。

B：Yes, thank you. It's very delicious. Is it one of your famous dishes?

B：谢谢，味道太香了。这是中国的名菜吗？

A：Yes. Do you like Chinese food?

A：是的。你喜欢中国菜吗？

B：Sure. It's marvelous. I think Chinese food is the best in the world.

B：当然，味道好极了。我觉得中国菜是世界上最美食品。

A：Would you like some Chinese Baijiu?

A：喝点中国的白酒行吗？

B：Yes, just a little. Thank you again for preparing such a splendid banquet.

B：好的，就一点。谢谢你为我准备的丰盛的晚宴。

A：It's our pleasure. What has impressed you most during your stay in Hangzhou?

A：很荣幸。杭州留给你最深的印象是什么？

B：The history of Southern Song Dynasty, the hospitality of Hangzhou people and the food. Of course, our successful business talk has impressed me most.

B：南宋历史，杭州人民的好客，还有中国菜。当然印象最深的还是我们这次成功的商务会晤。

A：I propose a toast to our friendship! Cheers!

A：我提议为我们的友谊举杯！干杯！

B：Cheers!

B：干杯！

◇ **Dialogue Ⅴ**

这段对话发生在杭州萧山机场，杭州红粉服饰有限公司外贸业务员 Dicky 送前来考察的美国斯特朗服装有限公司采购部经理 Betty 上飞机。

A：杭州红粉服饰有限公司外贸业务员 Dicky

B：美国斯特朗服装有限公司采购部经理 Betty

A：Miss Betty, it's all set. You have got one piece of checked luggage and a carry-on luggage, haven't you?

A：Betty 女士，手续都办理好了。您总共就一件托运行李和这一件随身行李吧？

B：Yes, that's right. Thank you very much, Dicky. You are so considerate and helpful that not only arranged the wonderful trip, but also come to see me off today. I really appreciate it.

B：对，没错。太感谢您了，Dicky。您真的很周到，不但把这一周的行程安排得那么好，今天还专程来为我送行。真的很感激。

A：Don't mention it. It's a pleasure to help you. We will miss you and hope to keep in touch in the future.

A：不客气，很高兴您能满意。我们会想念您的，也希望今后保持联系。

B：Of course. I also look forward to your visit to our annual meeting next October.

B：当然。也期待你明年 10 月能来美国参加我们公司的年会。

A：Sure, thank you. I will. It's time to get aboard.

A：好，谢谢，我一定来。该登机了。

B：Yeah, I should go, too. Thank you very much. Bye!

B：是啊，我该走了。非常感谢！再见！。

A：Goodbye! Wish you a good journey.

A：再见！旅途愉快！

Roles Simulation

Suppose you are one of them in the following conversation. Try to read aloud and practice the underlined sentences.

这段对话发生在杭州红粉服饰有限公司外贸业务员 Dicky 与美国斯特朗服装有限公司采购部经理 Betty 之间。作为东道主的杭州红粉服饰有限公司委托 Dicky 代表公司利用谈判间隙带着 Betty 购物观光。

A：美国斯特朗服装有限公司采购部经理 Betty

B：杭州红粉服饰有限公司外贸业务员 Dicky

B：Betty，shall we walk around the city and do some shopping this afternoon？

B：Betty，今天下午我们去逛街购物好吗？

A：Great! I'd just like to go shopping to purchase some souvenirs.

A：好呀，正好想买点纪念品回去。

B：Do you have anything special in mind that you would like to buy?

B：有特别想买的东西吗？

A：I got no idea.

A：还没想好。

B：How much were you thinking of spending to buy souvenirs?

B：大概准备花多少钱买纪念品呢？

A：Somewhere around 100 dollars.

A：一百美元左右。

B：Let's go to E-fang Block. <u>E-fang Street is located in Hangzhou Shangcheng Street. E-fang Block is an ancient street with a long history and profound cultural heritage. It is the cultural center and economic and trade center of the Southern Song Dynasty. There are special snacks, antiques, calligraphy and painting here.</u>

B：去河坊街吧。<u>河坊街位于杭州市上城区，是一条有着悠久历史和深厚文化底蕴的古街，是南宋的文化中心和经贸中心，这里有特色小吃和古玩字画。</u>

A：I'm sure that I can buy a real bargain souvenir there.

A：相信在那一定能买到真正物美价廉的纪念品。

B：How about a dark-red enameled pottery? <u>The noble and vulgar purple sand teapot is a perfect combination of art and practicality. The purple sand teapot making tea integrates Zen culture, which is a work of art worth collecting. The color and fragrance of the tea are all contained.</u>

B：买个紫砂材质的陶器怎么样？<u>高贵不俗的紫砂壶是艺术与实用的完美结合，融入禅文化的紫砂壶泡茶色、香、味皆蕴，是值得收藏的一件艺术品。</u>

A：Yes, I heard of it. But I know nothing about it. Can you help me find a good one?

A：可以。但我一点都不懂，你帮我挑一个吧。

B：My pleasure.

B：乐意为您服务。

(Two hours later, they walked to the West Lake.)

(两小时后，他们来到了西湖。)

A：What a charming place! <u>It may be the best-known scenic spot in the city.</u>

A：多么迷人的地方啊！<u>也许，这是本市最著名的风景区吧！</u>

B：This is the West Lake well-known in the world.

B：这就是闻名于世的西湖。

A：A lot of people are boating on the lake.

A：许多人正在湖里划船。

B：<u>All of the flowers are in full bloom, and the reflection of the pagoda in the lake looks very nice. It is the most beautiful season here.</u>

B：<u>所有的花朵都开了，倒影在湖心的塔影显得漂亮极了。现在是这里最漂亮的季节。</u>

A：<u>One feels rather relaxed when sitting here, enjoying the beauty of the scenery, in the shade of these ancient trees.</u>

A：<u>坐在这些古树的林荫下，欣赏着这美丽的景色，真使人感到轻松愉快。</u>

B：<u>Would you like to take a photo to leave the unforgettable time?</u>

B：<u>拍个照片记下这难以忘怀的时刻，好吗？</u>

A：Of course.

A：当然可以。

B：Look at those mountains! They are called Wu Mountain. <u>Let's climb up the hill. There is a beautiful garden over the hill and we can get a fine view of the city from the top.</u>

B：看那边的山！这是吴山。<u>我们到山顶上去吧。那儿有一个美丽的花园，在上面可以一览杭城的美景。</u>

A：OK. Let's go there.

A：好的。我们上去吧。

B：Come on, Betty.

B：好，走吧，Betty。

Typical Expressions

1. Making reservation 订酒店

A：I'd like to book a standard room. 我想订个标间。

B：I'm afraid the double room is fully booked. 对不起，双人间已经满了。

A：Is there a single room still available? 请问有单人间吗？

B：Would you like to make a booking? 请问您要定房吗？Can I have your name, please? 姓名？

A：The booking is for one passenger on Xiamen Airways Flight DF5539 at 09:20 am. 给厦门航空 9 点 20 分航班 DF5539 客人订个房间。

B：Could you send an e-mail to confirm the booking? 请将订房信息发个邮件过来好吗？

2. Making invitations 口头商务邀请

A：Would you like to go sightseeing around Hangzhou? 你想去杭州街上逛逛吗？

B：Yes, I'd be delighted. 可以呀，非常高兴。/ It's very kind of you.你太好了。/ Thanks, that sounds great. 谢谢，听起不错。

A：We're thinking of visiting Lingyin Temple, are you free? 我们在考虑去灵隐寺，你有空一起去吗？

B：Good idea, what time shall we meet? 好主意，我们什么时候碰面？/ Yes, I'd like to join you. 好的，非常乐意。/ Thanks, but I have a meeting. 谢谢，我还有个会。

A：Have you tried the Chinese dishes? 吃过中国菜吗？

B：I'm afraid I don't like spicy food. 我恐怕不喜欢辛辣食物。

A：Can I offer you a cup of coffee? 我给你倒杯咖啡怎么样？

B：Yes, Please. 可以的。/ A glass of mineral water or a cup of black tea, please. 请给我矿泉水或者红茶吧。/ I'm sorry. I think I prefer hot water. 对不起，来杯热水吧。

3. Expressions for correspondence 商务信函常用语

(1) Informal and neutral writing 非正式信函

Hi/Hello. Best wishes/regards(e-mail) 你好，祝安好！(电邮)

Dear John, Best wishes/regards(letter) 约翰，祝安好！(书面信函)

(2) Formal letters 正式信函

Dear Mr. (Brown) / Dear Sir/Madam 尊敬的(布朗)先生/先生/女士

Yours sincerely / Yours faithfully 您忠实的朋友

4. Expressions for making invitations 书面邀请常用语

I am writing to invite you to visit our plant. 我写信邀请您来参观我们的工厂。

We would be delighted if you could come to Hangzhou. 如果您能来杭州，我们将会非常高兴。

5. It's a pleasure to meet friends coming from afar. 有朋自远方来，不亦乐乎！

Anyhow, I think you'd like to freshen up a bit and take a rest to overcome the jet lag. We'd better start for the hotel now. 尽管如此，我想你们一定愿意稍加休息，恢复一下，调整时差。我们现在最好还是去旅馆吧。

6. Is there any place you'd like to visit in particular? I could help you arrange that.您有没有什么地方想要参观的？我可以为您安排。

If memory serves, it has been almost two years since our last meeting. 如果没记错，我们上次见面到现在差不多有两年了吧。

7. Let's make it Thursday afternoon, say, 3 o'clock? 我们定在星期四下午，三点钟，怎么样？

He shall be delighted to see you in your office this afternoon if this is convenient for you. 如果您方便的话，他很愿意今天下午在您办公室与您见面。

I wonder whether we could reschedule the Tuesday appointment from 8 a.m. to 9 a.m.? 不知能否将下周二早上 8 点的会面改为早上 9 点？

8. Hopefully I haven't messed up your arrangements too much. 但愿我没有过多地打乱您的安排。

9. On behalf of our corporation, I want to extend a warm welcome to you and thank you for your friendly cooperation you have shown us over the years. 我谨代表我们公司，向你们表示热烈的欢迎，并感谢你们过去几年来的友好合作。

10. May I propose a toast to the success of our negotiations(to our friendship and cooperation)? Cheer! 我提议，为我们谈判的成功(为我们的友谊和合作)，碰杯！

Foreign Trade Letters

◇ Sample Ⅰ　A thanks letter

After meeting, normally we should send an e-mail to say thanks!

Dear Kristin,

Thanks for coming and nice to meet you at our factory, thanks!

Have a nice trip in China.

Best regards!

Martin

◇ Sample Ⅱ　A feedback letter

After meeting, we must recap a meeting memo.

发件人: XXXXXX

Date: 2021 年 10 月 23 日周二 下午 6：50

Subject：ZONGSEN 10/23 meeting recap

To:XXXXXX

Cc: XXXXXX

Hi, everyone. Please see attached for the recap from my meeting with ZONGSEN today.

All updates are in BLUE.

Best regards,

XXXXXX

Ronny Kobo, Director of Production

Attachment: the recap

　　　　RONNY KOBO

　　　　ZONGSEN-October 23,2021

General Points:

- Invoices
  - ➢ Production and development invoices must be separate
  - ➢ Invoices must be for one season only—cannot mix season on one invoice
  - ➢ Production invoices-PPS, TOP samples, bulk production send to Kristin's attn
  - ➢ Design/development invoices-sample yardage/trims, proto samples, sales samples send to Kaitlyn/Jessica attn
- Sample sending
  - ➢ Email alert of all packages sent with awb#
  - ➢ All samples to be sent to Caroline's attn
  - ➢ All samples must be tagged with supplier ID, style# and sample type
  - ➢ All PPS and TOP samples must be sent with factory spec
  - ➢ PPS must be in bulk fabric/yarn and trims when possible
  - ➢ TOP samples must be sent in correct packaging, labeling and hangtag
    - ■ Sent manual for care/content info, label/hangtag placement, sticker and poly bag info;
    - ■ Sent ACT ONE PO template to book mainlabels, size/COO labels and hangtags
  - ➢ All sales sample must have Ronny Kobo main labels sewn in
- Nominated Fabrics
  - ➢ We sometimes need nominate a fabric source that is outside of China. How is fabric imported?
  - ➢ Walk through the import process, duty, VAT and other costs
  - ➢ What is refundable after shipping finished goods? How is this cost determined in the cost sheet?
- WIP Charts
  - ➢ To be updated and sent weekly, any day Monday-Thursday is ok
- Shipping Quantity Tolerance
  - ➢ +/-10% for < 100pcs
  - ➢ +/-5% for 100-300pcs
  - ➢ +/-3% for 300pcs+

■ **Notes：**

recap: 会议记录要点

◇ Sample Ⅲ An enquiry letter

To: Helen

Fm: Martin

CC: XXXX

RE: Inquiry for XXXX

My name is Martin and I'm contacting you on behalf of [company name]. I would like to inquire about one of your products, [product name]. I would like to have an idea about the different models, features, and options. Also please tell me about the available colors, prices, bulk order discounts, warranty, delivery, and credit payment option. I appreciate if you could also enlighten me on the other alternatives that might also suit our needs. I look forward to your response.

Thank you!

Best regards!

Martin

■ Notes：

Enquiry Letter Tips

(1) Begin your letter by stating who you are and giving your status or position.

(2) Clearly state what it is that you are inquiring about and what you would like the recipient of your letter to do.

(3) You might want to briefly explain the purpose of your letter or what you hope to accomplish.

(4) If appropriate, consider mentioning the letter recipient's qualifications for responding to your inquiry.

(5) Include the date by which you need the information, services, etc., that you are requesting, and indicate that you await the reader's response.

(6) Thank the person for his /her time.

◇ Sample Ⅳ　An enquiry letter

Dom　XXXXXX@126.com

发送至 XXXXX

Hi, Martin

We have a small knitted order enquiry in 100% merino wool(美诺羊毛) in 12 gg or 100% Cashmere(羊绒) 3 gauges hats/Beanie.

Please see attached. I do now know if you have interest to help to estimate a quote or not.

For Cashmere, only 100 pcs. For Merino Wool, could be 200 pcs.

Please advise if you have interest and make us a quote.

Regards,

Dom

■ Notes：

　　GG(gauge)：指排针标准。通常指毛制品编织时针板上每英寸排针的数目。数字越大针越细，用的纱线越细，织出来的毛制品越密；相反，数字越小针越粗，用的纱线越粗，织物密度越稀。比如 5 针即每英寸排 5 根针；5GG 织物比 7GG 织物更松散。

◇ Sample Ⅴ　Reply to sample Ⅳ

Hi, Dom

Thanks for your inquiry and support.

We are willing to make these orders and will offer our best price by tomorrow.

If you have any other inquiry. Pls don't hesitate to send us. Thanks.

Best regards,

Hedy

Oral Practice

Forming a group of two or three partners. Try to work out a dialogue related to the theme of this unit and then perform it in class.

Self-Assessment

Evaluate your practice by marking the following each corresponding item and work out the total scores.

考核情境(分值)	考核要求(分值)	得 分
语音语调(20分)	发音准确,声音清晰(10分)	
	语调自然,语速自然流畅(10分)	
内容(30分)	内容完整(10分)	
	内容表述符合给定情境(10分)	
	前后内容逻辑一致,内容之间无矛盾(10分)	
语言描述(30分)	用词规范、准确,无语法错误(20分)	
	表述无歧义(10分)	
其他(20分)	无雷同状况(10分)	
	无严重偏题(10分)	
合 计(100分)		

Task 3　Products and Service

知识目标

◆ 学习国际贸易中产品与服务的介绍
◆ 学习国际贸易中产品与服务介绍的途径
◆ 学习相关商务沟通的口头英文表达方式
◆ 学习相关商务沟通的书面英文表达技巧

能力目标

◆ 能够准备产品与服务介绍材料
◆ 能够顺利开展产品与服务介绍活动
◆ 能够使用双语顺畅地进行商务口头沟通
◆ 能够处理产品与服务介绍往来信函

Basic Knowledge Review

1. What does negotiation on products and service involve?

During the stage of negotiation on products and service, exporters always describe their specific products and services. Direct communication is one of the most persuasive things there is. Exporters should fully explain the concept for their business, along with all aspects of purchasing, manufacturing, packaging, and distribution. Let's overview the details as follows:

(1) Identify the name of the company or organization.

(2) Tell about the company or organization. Mention how it was started, how long you've been in business, your mission or business objective or goals, and so forth.

(3) If appropriate, identify the product(s) or service(s) you provide, and identify the benefits of buying or using these products or services. Tell why they are better than those of the competition; how they will save the reader time/money, make his/her quality of life better, or help him/her to accomplish a certain goal; and so on. In short, tell the reader why he/she must have the product or service you offer.

(4) Invite the person to an open house, grand opening, sale, etc., if applicable.

2. How do we achieve the goal of negotiation on products and service successfully?

(1) Before we go on negotiation, we should practice the following business principles.

Put yourself in your customer's shoes. Too often, entrepreneurs fall in love with their product or service and forget that it is the customer's needs, not their own, that they must satisfy. Step back from your daily operations and carefully scrutinize what your customers really want. Suppose you own a pizza parlor. Sure, customers come into your pizza place for food. But is food all they want? What could make them come back again and again and ignore your competition? The answer might be quality, convenience, reliability, friendliness, cleanliness, courtesy or customer service.

Remember, price is never the only reason people buy. If your competition is beating you on pricing because they are larger, you have to find another sales feature that addresses the customer's needs and then build your sales and promotional efforts around that feature.

(2) Know what motivates your customers' behaviour and buying decisions. Effective marketing requires you to be an amateur psychologist. You need to know what drives and motivates customers. Go beyond the traditional customer demographics, such as age, gender, race, income and geographic location, that most businesses collect to analyze their sales trends. For our pizza shop example, it is not enough to know that 75 percent of your customers are in the 18-to-25 age range. You need to look at their motives for buying pizza-taste, peer pressure, convenience and so on. Cosmetics and liquor companies are great examples of industries that know the value of psychologically oriented promotion. People buy these products based on their desires (for pretty women, luxury, glamour and so on), not on their needs.

(3) Uncover the real reasons customers buy your product instead of a competitor's. As business grows, you'll be able to ask your best source of information: your customers. For example, the pizza entrepreneur could ask them why they like his pizza over others, plus ask them to rate the importance of the features he offers, such as taste, size, ingredients, atmosphere and service. You will be surprised how honest people are when you ask how you can improve your service.

After we know the demand of customers, it is a key step to clarify the selling points of products and service clearly to arouse buyers to make an immediate order. Here are some tips to help us to do that.

(4) Don't be wordy. Generally, a product description shouldn't be more than 200 words; this keeps it above the fold of the webpage. If you're selling a service (particularly one where there is a process involved), you've got more leeway. Remember that brevity is the soul of wit, and also the soul of business. Introduction letter templates and products catalogues need to communicate your product's or service's most appealing features in a short message. Use short, active

sentences that are easy to read and remember.

(5) Be creative. Those who've been around the business block a few times likely have heard of the USP, or the Unique Selling Proposition. Basically, it's a sentence that tells everybody why you're different from the rest. This might be harder than you think—assuming you're selling widgets, there are hundreds of others out there who are likely doing the same thing if it's at all profitable. Why is your business different from everybody else's?

This is a good one to sit down and think about for a while. If you're swimming in a sea of competitors who are all selling similar things, why should the customer choose you? Distill that uniqueness and put it in your product copy. You'll be amazed at the magic.

(6) Make it easy for your customers to learn more. Provide free trials, downloads, product videos, and demos.

3. Introduction to Sampling Letter
(1) How to send export samples to international buyers?
Sampling is the first step in the process of exporting product. When exporters deal with the foreign buyers, he should always be prepared in order to provide samples of his products. Promotional measures are beneficial for a healthy start. It thus depends on him to provide his product samples whether they are free of cost or on a minimum charge.

Samples having permanent marking as "sample not for sale" are allowed freely for export without any limit. However, in such cases where indelible marking is not available, the samples may be allowed for a value not exceeding US $10,000, per consignment. Sampling plays a vital role in getting orders from the buyer.

It is a representation of the manufacturer's or exporter's potential to deliver desired outputs based on the details and information specified by the buyer. Besides this, sampling also gives an idea regarding the time and cost that the manufacturer of the garments will require completing the whole order, and delivering the products in time.

(2) Steps to send export samples to international buyers.
For export of sample products which are restricted for export as mentioned in the HS Code, an application may be made to the office of Director General of Foreign Trade (DGFT).

There is a checklist, one need to understand and move step by step:
✓   Import-Export Certificate (IEC)
You as a businessman, first & foremost needs to have an IEC—registration for Export Import

under DGFT.

✓ Export items

You can also send / export freely "exportable" items on free of cost basis for export promotion subject to an annual limit set to a license (IEC). Well for a first timer, there should not be a problem. Later on you may consult your Chartered Accountant for this financial clause.

✓ H.S. Code

You must have the product information to the extent. You and your buyer are clear on the exact H.S. Code and specifications of the product.

✓ Product Knowledge

It is very important that you have detailed product knowledge that you are going to send to a buyer. For example: It should not be that you are talking of "Sour Lime" and the buyer in Dubai understands it as "Eureka Lemon".

✓ Packaging

Now is the packing time. Your product must be clean, without any dust, without any patches, and should be brand new.

✓ Invoice

Take a format from Digital Exim Documents. Invoice must have clear mention "SAMPLE— NOT FOR SALE". Invoice must be printed & not handwritten.

✓ GST (Goods and Services Tax)

GST originates from India. Generally speaking, for the samples there is no GST. However, if you have bought your free sample from a manufacturer then GST is paid to the supplier / manufacturer.

✓ Dispatch

Send your samples as a GIFT package. This will create a lasting impression on your buyer. As you know, the first impression is last impression which is very important.

✓ Courier

Send the samples through postal channel, courier or airmail. Take a proper receipt for the payment of whichever channel you choose and keep a track of it. Examples of courier companies: DHL, TNT, FedEx, Blue Dart, Aramex, Speed Post etc.

✓ After sample is received

Once the sample received by the buyer, give a courtesy call to the prospective buyer. Ask him how the product is, does it match his expectations. Talk about final pricing and close the order.

(3) Types of samples used in the garment export industry

A sample of a garment can either make or break the future of getting orders from buyers. A good quality sample can invite more business and buyers to manufacture garments. It also gives the manufacturer the opportunity to look for options to source fabrics, trims, and other garment accessories necessary for the design at a cheaper price but of better quality, which further helps in costing. It helps in optimizing the process parameter for mass producing and also can aid in eliminating bottlenecks.

On the other hand, buyers can keep a check on the production of garments, the designs, and styles that are required, which are generally outsourced to manufacturers situated at a far distance. Holding a physical sample of a garment helps the buyer check the look, feel, fall, colors and shades, fitting, and the pattern of a certain style.

Usually, there are several types of samples used in the garment export industry.

✓ Promotional samples or salesmen samples

Promotional samples or salesmen samples are developed to procure orders from retailers. These are good quality samples, which use actual accessories and fabrics. The buyer has to pay for these kinds of samples.

✓ Proto samples

Proto samples are made after receiving the order sheets, and are the very initial sets of samples sent to the buyer. The design and style of a garment are communicated; the fit and fabric detailing are not given so much attention. Once the proto samples are approved, fit samples are developed.

✓ Fit samples

Fit samples are designed to check the measurements, fit, and the style. Details of construction of the garment and standards needed to be maintained by the manufacturer are maintained in these samples. Certain times these samples are tried on models to check the fitting and the fall of the garment.

✓ Size samples

Size samples are sent to the buyers once the fitting is approved in small, medium, large and other sizes as per the requirement. Mass cutting of fabrics begins only once a final approval of all the sizes is received.

✓ Pre-production samples

Pre-production samples are the samples, which are made in the production department once all the above sampling approvals are received. They are made with the actual fabrics, trimmings, and accessories that will be used in the future placed order. After a full and final approval is received from the buyer, the actual production can proceed.

✓   Top of production samples

Top of production samples are the samples that are picked up in between, once the manufacturing process has begun. All buyers do not demand for top of production samples, but some do, to make sure that the garments being manufactured are as per the decided specifications.

✓   Shipment samples

Shipment samples are sent to the buyer when the completed garments are packed and are ready to be shipped. These samples are sent in cases when the garments are directly delivered at stores or warehouses of the buyers, to view the final product and its packaging.

Task Descriptions

为了让客户更多地了解产品与服务，大部分出口商会主动邀请客户参观工厂以增进与客户彼此间的信任，促成客户下单。为了让美国斯特朗服装有限公司采购部经理 Betty 更多地了解杭州红粉服饰有限公司生产能力和产品质量。

Products and Service

接下来的对话就是围绕杭州红粉服饰有限公司外贸业务员 Dicky 带领美国斯特朗服装有限公司采购部经理 Betty 参观工厂的一系列活动展开的。

Typical Dialogues

◇ **Dialogue** Ⅰ

这段对话是在杭州红粉服饰有限公司外贸业务员 Dicky 邀请美国斯特朗服装有限公司采购部经理 Betty 参观工厂时发生的。

A：杭州红粉服饰有限公司外贸业务员 Dicky

B：美国斯特朗服装有限公司采购部经理 Betty

A：Betty, what do you think of visiting our plant? It's better way to know our products.

A：Betty, 想不想到我们工厂去转转呢？看看工厂，您就会更了解我们的产品。

B：That's a good idea. If it wouldn't take too long to arrange.

B：要是很快能安排的话，这倒是好主意。

A：No, It won't. I'll arrange it for you.

A：不会的，我来给您安排一下。

(Ten minutes later)

(十分钟后)

A：We can arrange tomorrow's trip. What do you think about it?

A：我们可以安排在明天参观。您看可以吗？

B：That's great! I'm really looking forward to this.

B：太棒了！真的非常期待！

A：It will take about four hours. I will accompany you on your tour.

A：这次参观前后大概需要四个小时。我全程作陪。

B：It's very kind of you to do that. Will you show me around in the factory?

B：太好了。您带我参观工厂吗？

A：No. The manager of the plant will go with us and answer any questions at any time.

A：不。工厂经理会全程陪我们一起随时解答您的疑问。

B：Thank you for your consideration.

B：谢谢您考虑得这么周全。

◇ Dialogue Ⅱ

这段对话的内容是：杭州红粉服饰有限公司外贸业务员 Dicky 与美国斯特朗服装有限公司采购部经理 Betty 谈论产品的卖点问题。

A：杭州红粉服饰有限公司外贸业务员 Dicky

B：美国斯特朗服装有限公司采购部经理 Betty

A：Good morning, welcome to our showroom. What particular kind of products are you interested in?

A：早上好，欢迎参观陈列室。有没有特别感兴趣的？

B：How is the product selling?

B：产品卖得怎么样？

A：It's selling well.

A：卖得不错。

B：What are the selling points of your product?

B：卖点在哪？

A：Compared with competing products, ours is more modern and fashionable.

A：跟其他同类相比，我们家的衣服更现代更时尚。

B：I'd like to take these catalogs back to the hotel. And I want the price lists as well. May I?

B：我想把这些目录和价目单带回酒店，可以吗？

A：Sure. Go right ahead. Please take whatever you like.

A：当然可以。

B：Well, may I have this sample free of charge?

B：样品可以免费送我吗？

A：Of course, you can.

A：当然可以。

◇ Dialogue Ⅲ

这段对话的内容是：杭州红粉服饰有限公司外贸业务员 Dicky 向美国斯特朗服装有限公司采购部经理 Betty 介绍公司产品。

A：杭州红粉服饰有限公司外贸业务员 Dicky

B：美国斯特朗服装有限公司采购部经理 Betty

A：Miss Betty, these dresses are the main products of ours, and they are the latest models designed by us.

A：Betty 女士，这些连衣裙是我们的主打产品，都是我们自行设计的最新款式。

B：Wow, how nice. This sleeveless style with V neck and two-side pockets is the popular element this year.

B：嗯嗯，真的很不错。无袖 V 领带两个侧口袋会是今年的流行元素。

A：The models here on display in this showroom are just the newly-designed as well as some classic styles. We still have many other styles which you can find in this catalogue.

A：是啊，这个展室陈列的只是今年新开发的款式和一些常青(经典)款，我们还有很多其他的款式，您都可以在这本产品目录里找到。

B：OK. Will you show me the price list?

B：好的，能给我一份价目表吗？

A：Sure. Here you are. By the way, you can take each from this series as a sample for free, but you have to pay for these items with a sample discount of twenty percent since it's our company's policy to charge for any samples over five U.S. dollar.

A：当然。给您。对了，这个系列您可以免费各带一件作为样品，但因为这些款式成本已超过 5 美元了，所以依照本公司的规定是需要收费的。但每件可以打八折。

B：OK, thank you.

B：好的，谢谢。

◇ Dialogue Ⅳ

这段对话的内容是：杭州红粉服饰有限公司外贸业务员 Dicky 与美国斯特朗服装有限公司采购部经理 Betty 谈论产品的款式问题。

A：杭州红粉服饰有限公司外贸业务员 Dicky

B：美国斯特朗服装有限公司采购部经理 Betty

A：These sleeveless dresses with V neck are in large demand, and are sold out in a lot of shopping malls. Many more orders have been added this month.

A：这款无袖 V 领连衣裙需求量很大，很多商场都卖到脱销，这个月补增了很多订单。

B：Mm-hmm, this sleeveless V-neck design combined with an elastic waist can not only help one show off a good shape, but also be graceful.

B：嗯嗯，无袖 V 领加上束腰的设计不但能勾勒出凹凸有致的身段，还显得非常有气质。

A：Yes, with high heels, they can be more elegant and intellectual. Actually, the floral design of this round neck is also with the designer-look.

A：是呢，搭配高跟鞋，更加优雅知性。其实这款圆领的碎花款也很有设计感。

B：This round-neck model seems more lovely and lively. Being casual, it is more suitable for the seaside holiday. However, the V-neck style suits all kinds of occasions. Wearing a V-neck dress, one will seem to be taller than with a round-neck dress. With a V neck, the face can be seemed smaller. Besides, the V neck helps stretch the neck line and show the sexy clavicle at the same time.

B：圆领这款更显可爱活泼，比较偏休闲，更适合去海边度假。而 V 领的款式就很适合各种场合的穿搭。穿 V 领似乎比穿圆领更让人显得高挑。V 领会显得脸小，拉伸脖子线条的同时还能露出性感的锁骨。

A：Yes, I thought this kind of sleeveless style would only be favored by girls under 30, but I didn't expect it would be more popular among women in their 30s and 40s. It is said that they prefer to wear a short suit or cardigan over this dress.

A：是啊，本以为这种无袖的款式只会受到 30 岁以下的小姑娘的青睐，没想到它在三四十岁的女性中更受欢迎。听说她们更喜欢在这款连衣裙外搭一件短西装或短开衫。

B：Yeah, this kind of wearing will be steady and generous.

B：没错，这样的搭配确实会更显稳重大方。

◇ **Dialogue V**

这段对话的内容是：杭州红粉服饰有限公司外贸业务员 Dicky 带美国斯特朗服装有限公司采购部经理 Betty 参观工厂。

A：杭州红粉服饰有限公司外贸业务员 Dicky

B：美国斯特朗服装有限公司采购部经理 Betty

A：Miss Betty, today let me show you around the whole workshop.

A：Betty 女士，今天我带您参观整个车间。

B：OK, thank you.

B：好的，谢谢。

A：This is our latest production line imported from Italy. We have a production capacity of 10,000 pieces per day.

A：这是我们从意大利新引进的生产线，日生产能力 10,000 件。

B：Wow, how do you control the quality?

B：喔，你们是怎么来控制质量的呢？

A：All products are under our supervision. They are to pass strict inspection before going out, so there are a total of six checks in the entire manufacturing process.

A：所有产品都在我们的全程监督之下。它们都要通过严格的检验才能出厂。整个制造过程中总共要经过六道检查。

B：Do the staffs have to do the shift work?

B：工作人员需要轮班吗？

A：As the line is fully automated, less staffs are needed than ever before. However, as the line is under 24-hour operation, the staffs still have to work shifts.

A：由于生产线是完全自动化的，所需的员工比以往任何时候都要少，但是由于生产线是24 小时运作的，所以工作人员仍然需要轮班工作。

B：That's good. Your workshop seems so tidy, and everything is in order.

B：很好。你的车间很整洁，一切都井然有序。

A：Since it is the peak season now, we are all booked up, but the quality is always of high priority. We will also ensure that our staffs have adequate time to rest.

A：由于现在是旺季，我们的生产档期都已排满了。但质量永远至上，我们也会确保员工有正常和充足的休息时间。

B：Thank you, and I'm so pleased to hear that.

B：谢谢，我很高兴听到您这么说。

🎎 Roles Simulation

Suppose you are one of them in the following conversation. Try to read aloud and practice the underlined sentences.

　　这段对话的内容是：杭州红粉服饰有限公司外贸业务员 Dicky 带美国斯特朗服装有限公司采购部经理 Betty 参观工厂。

A：杭州红粉服饰有限公司外贸业务员 Dicky

B：美国斯特朗服装有限公司采购部经理 Betty

A：Welcome to our factory.

A：欢迎来工厂参观。

B：I've been looking forward to visiting your factory.

B：我一直想参观贵工厂。

A：I'll show you around and explain the operations as we go along. Actually，you'll know our products better after the visit.

A：我们边走边谈，我一会向你介绍一下生产过程。事实上，参观完工厂之后你将会对我们的产品有更好的了解。

B：That'll be most helpful. How large is the plant？

B：太有帮助了。工厂多大？

A：It covers an area of 55,000 square meters.

A：工厂总面积 55,000 平方米。

B：It's much larger than I expected. When was it established？

B：比我预想的要大。工厂什么时候建的？

A：In the early 1970s. We'll soon be celebrating the 51st anniversary.

A：20 世纪 70 年代。我们很快要庆祝建厂 51 周年了。

B：Congratulations.

B：祝贺！

A：Thank you.

A：谢谢。

B：How large is the machine shop？

B：机器间多大？

A：Its total area is 3,000 square meters.

A：3,000 平方米。

B：Do we have to wear the helmets？

B：我们要戴头盔吗?

A：You'd better protect your heads.When in the workshop，please don't cross the white lines.

A：你最好保护一下头部。到了车间，不要跨过白线。

B：Is the production line fully automatic？

B：生产线是全自动的吗？

A：Well, not fully automatic. <u>Maybe we could start with the Design Department. And then we could look at the production line.</u>

A：不全是自动的。<u>也许我们可以从设计部开始，在那我们看到整个生产线。</u>

B：<u>How much do you spend on design development every year？</u>

B：<u>每年在设计部开支多少？</u>

A：About 3％～4％ of the gross sales.

A：毛销售总额的 3％～4％。

B：That's fine. Could I have a look over the manufacturing process？

B：不错。我能参观一下生产工序吗？

A：Of course. This way，please.

A：当然可以，请这边走。

(They are on the way to the production line.)

(他们在去生产线的路上。)

A：This is the assembly line. <u>These drawings on the wall are process sheets. They describe how each process goes on to the next.</u>

A：这是装配线。<u>墙上的这些图纸是生产流程图。从流程图上可以看出每道工序是如何到下一道工序的。</u>

B：How many workers are there on the lines？

B：生产线上有多少工人？

A：About 60. But we are running on two shifts. This assembly line was built recently. <u>Almost every process is computerized. What a worker does is usually to push buttons. Thus，the</u>

efficiency is greatly raised，and the intensity of labor is decreased.

A：大概 60 个。但我们是分两班工作的。<u>装配线最近刚建的。每道工艺都是计算机控制的。</u><u>工人只需要按按钮。</u>这样既提高了工作效率又降低了工作强度。

B：What kind of quality control do you have？

B：你们采用的是什么质量控制？

A：All products have to pass strict inspection before they go out. All products have to go through five checks in the whole process. <u>We believe that the quality is the soul of an enterprise.</u> <u>Therefore, we always put quality as the first consideration.</u>

A：所有的产品出厂前都经过严格检验的。整个的检验过程一共有 5 道工序。<u>我们坚信质量是一个企业的灵魂。因此，我们总是把质量放在第一位。</u>

B：Yes<u>, quality is even more important than quantity.</u>

B：是的，<u>质量比数量更重要。</u>

A：That's right.

A：对的。

B：This visit gave me a good picture of your product areas. I hope my visit does not cause you too much trouble.

B：通过这次参观我对产品有了更好地了解，但愿没给你们添太多麻烦。

A：No trouble at all. I'm pleased you found it helpful. What's your general impression, may I ask？

A：不麻烦。很高兴你认为参观是有帮助的。可以问一下，您对我们工厂的总体印象吗？

B：I'm impressed by your approach to business. In addition, you have fine facilities and efficient people. The product gives you an edge over your competitors, I guess.

B：打动我的是你们的商业战略、精良的设备与高效率的工人。我猜，产品是你们战胜对手的法宝。

A：Yes. No one can match us as far as quality is concerned.

A：是的。质量上没有谁能比得过我们。

B：Could you give me some brochures for your product？ And the price if possible.

B：产品宣传册给我一些，行吗？如果可能，产品的价格也附上。

A：Certainly. Shall we rest a while and have a cup of tea before going around？

A：可以。歇歇，喝杯茶再转吧。

B：That's fine.

B：好的。

🐾 *Typical Expressions*

1. When we arrive at the plant, I will give you a tour of the facilities. 到了工厂，我会请您参观设备。

2. Here is a brochure outlining the history and products of our company. 这里有一本关于我们公司历史和产品的小册子。

3. This is our office block. We have all the administrative departments here. Down there is the Research and Development Department. 这是我们公司的办公大楼。所有的行政部门都在这里。下边是研发部。

4. The entire manufacturing process is under our supervision. 我们负责监督全部的生产过程。

5. Our factory has a production capacity of 100 dozen per day. 我们工厂的生产能力是每天100 打。

6. Since it is peak season now, our factory is working at full capacity (we are all booked up). 因为现在是旺季，我们工厂的生产期已经排满了。

7. This product comes in three different colors: black, red and brown. 这种产品有三种不同颜色：黑色、红色、褐色。

8. The FGB900 is the best-selling product of its kind. FGB900 是在同类产品中销路最好的。

9. What's the material you used for this product? Is it made of goat leather? 这种产品的质料你们用的是什么？是山羊皮做的吗？

10. Do your products take longer to wear out than the others? 你们的产品会比其他产品耐用吗？

Foreign Trade Letters

◇ Sample Ⅰ   Proto sampling request

Hi Martin,

Hope you had a nice time for Chinese New Year and my most sincere wish for a very happy and prosperous New Year with loads of good things to you and your family!

As per our phone conversation, I am sending you again the original sample of GLM0386/CLA. Buyer wants absolutely the same quality and same knitting.

The duplications that Simi sent me were in Jersey and were in below quality,

50% viscose   26% polyamide   24% polyester

Can you confirm that original sample in above quality?

For us

Original sample is in 70% viscose   30% polyamide.

Now buyer wants exactly same as original—same quality/same knitting—kindly proceed and confirm the composition.

Take into account the changes—6 eyelets on each sleeve + same jewel on front.

As I told you, eyelets and jewel should be of good quality so that there are no problems with the tests.

Do not forget the bake neck tape.

Do not forget the hanger loops—transparent and sewn next to neckline at shoulder seam level.

Respect all comments made.

Based on above. Rush to send 2 samples top top urgently for our buyer.

Buyer will wait for these samples to issue the request for PPS and the order.

We rely on you and await your feedback.

Thanks to confirm that everything is clear.

BR

■ **Notes：**

1. Jersey：平纹针织
2. viscose：粘胶纤维
3. polyamide：聚酰胺，聚酰胺是用来生产尼龙或锦纶的原料。
4. polyester：聚酯纤维也称为涤纶。
5. the bake neck tape：后颈胶带
6. the hanger loops：吊带扣
7. PPS: Pre-production samples：产前样

◇ **Sample Ⅱ Salesman sample request**

Subject：SU-H19-PU16-LIF/SU 270001 URGENT SAMPLING REQUEST FOR WINTER 2021

Dear Martin,

Hope you are doing well.

Please note that SU buyer has selected your sample ref SU-H19-PU16-LIF/SU for their

WINTER/2021 collection.

Indeed, our client SU has selected this style and we have a firm order. We sent you color swatches last Friday thru DHL.

Please develop urgently the following salesmen samples, please follow below instructions:

STYLE /FINISHING/ KNITTING / QUALITY: Same as your sample.

COLOR: 2 COMBOS

LIGHT MARL GREY = I sent you a swatch for LIGHT MARL GREY but finally do not follow it. You can develop SMS as your previous sample in LIGHT MARL GREY.

PAON/PETROLE = as swatches sent. For this second combo, the body is PAON. For the text, letters T&G are to be in MILK and the rest letters are to be in PETROLE (O&E&T&H&E&R)

MEASUREMENTS: size 38/40 as your sample.

NUMBER OF SMS REQUIRED PER COMBO: 4PCS

TOTAL OF SMS REQUIRED: 8 pcs SMS + 1 technical sample in size 38/40 (Technical sample can be without embellishment, in available color but correct quality)

Thanks to attach hanger loops to your SMS.

DEADLINE: 20/12/2021 in France office

It is very important to receive your SMS on time for buyer's big presentation to their shops.

Indeed, all their shops are coming to the show and confirm their quantities during the show.

We count in you.

Pls confirm all clear to you.

Meilleures salutations / Best regards

Thi-Oanh CHANTHAVONG

Chef de produits

■ **Notes：**

1. SU：客户名称缩写

2. color swatches：颜色布片

3. Pls：Please

4. Thru: through

5. DHL：中外运敦豪速递公司

6. combo：组合

7. Thi-Oanh CHANTHAVONG：人名

8. chef de produits：(法语) 生产总监

9. count in：把……计算在内，在此的意思是依靠，指望的意思。

10. Meilleures salutations：(法语) 此致敬礼

11. ref: reference：参考

12. LIGHT MARL GREY：淡泥灰色

13. PAON/PETROLE：(法语) 孔雀蓝/石油蓝

14. SMS：salesman sample 销售样

15. MILK：奶白色

16. embellishment：图案排位

◇ Sample Ⅲ   Fit sample comment

DEAR SIMI,

PLEASE FIND HEREWITH NEW ORDER FOR WINTER 2021 COLLECTION.
-     ORDER LIF/10/2020
-     REF SU-H19-PU16-LIF/SU
-     QTY = 2540 PCS
-     COLOR = LIGHT MARL GREY + PAON
-     SHIPMENT DATE = 21/05/2019 – ETD VESSEL
-     TARGET PRICE = 7.40 USD

COMMENTS ON YOUR SMS IN LIGHT MARL GREY + PAON

SMS IN COMBO LIGHT MARL GREY = THEY ARE OK FOR STYLE, QUALITY, FINISHING AND COLOR.

SMS IN COMBO PAON = THEY ARE OK FOR STYLE, QUALITY, FINISHING BUT BACKGROUND COLOR IS NOT OK. THANKS TO SEND US URGENTLY LAB DIPS FOR COLOR PAON.

SIZE SPEC: PLEASE FIND HEREWITH SIZE SPEC TO BE FOLLOWED FOR BULK PROD.

THANKS TO SEND 1 TECH SAMPLE IN SIZE 38/40 IN ANY COLOR.

THIS REFERENCE WILL BE IN PROMOTION.

THANKS TO SEND 3 PHOTO SAMPLES IN SIZE 38/40 IN LIGHT MARL GREY FOR MARKETING PURPOSE.

WE NEED THESE PHOTO SAMPLES FOR FIRST WEEK OF MARCH DEADLINE.

THANKS TO CHECK THE PRICE WITH MAURICE DURING MEETING.

Meilleures salutations / Best regards

Thi-Oanh CHANTHAVONG

Chef de produits

◇ Sample Ⅳ Pre-production sample

Green light to production

发件人：XXXXXX

Date: 2021 年 8 月 20 日周一 下午 6:23

Subject: Comments CASS 115/OK PROD

To: XXXXXX

Dear Simi,

You can find herewith the remarks and the size spec for the style CASS 115.

Comments CASS 115; 3e received sample is size 38=S:(20/08/18)OK PROD

1. Please respect our comments!

2. The sleeve length is a total length with cuff, lapel (see the picture).

   63.5cm=59.5cm (sleeve) + 4.5cm (cuff)

3. Change the measure in red in the size spec.

4. The start of the waist dart = 28 cm from HSP (see the picture)

5. Please make the change from middle front. (see the picture)

6. The color print of the last sample, it's not OK! The mock up is OK.

OK PRODUCTION subject to respect our measures and comments.

Thank you for sending us a PPS size 38=SMALL for validation!

See attached file for detail PPS

XXXXXX

■ **Notes：**

1. green light to production：可以生产
2. herewith：随同此信(或书、文件)

   例：I enclose herewith a copy of the policy. 我随信附上一份保险单。
3. spec：规格
4. length：线段长度
5. cuff：袖口
6. lapel：(西服上衣或夹克上部胸前的)翻领
7. waist dart：腰省；腰褶
8. middle front：中锋
9. mock up：模型
10. PPS(Pre-production Sample)：产前样

◇ **Sample Ⅴ   Parcel sending information**

发送到 XXX

Dear XXX,

Today we send 1pc proto sample per version to you, via DLH#3788807886 please kindly note and check, many thanks!

Style no. MIBALSUEDE/PAT VERSION 11 pc+ VERSION 21 pc proto sample

Weight: version 1-364g/pc version 2-359g/pc

Composition: 80% acrylic 20% polyamide for yarn, 90% polyester 10% spandex for fabric

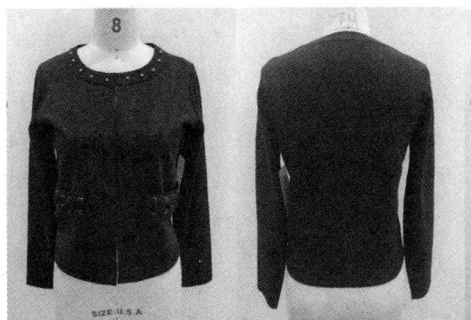

■ **Notes：**

1. acrylic：腈纶
2. polyamide：聚酰胺纤维俗称尼龙(Nylon)或锦纶
3. yarn：纱线
4. polyester：涤纶
5. spandex：氨纶
6. fabric：织物

Foreign Trade Documents

◇ **Sample** Ⅰ **Technical instruction-sketch**

ÐAMART		DESSIN TECHNIQUE / TECHNICAL DRAWING			IPC: **23716**	
		CREATED: 2014/1/8	UPDATED:		TECHNOLOGIST: C.DEBOSSCHERE	
FOURNISSEUR / SUPPLIER		DESIGNATION / DESCRIPTION		REFERENCE		
	SPRINGWAY				France REF: **23716**	
		pull fantaisie			UK REF:	
					Belgique REF:	
					Saison/Season:	**AH14** **WW14**

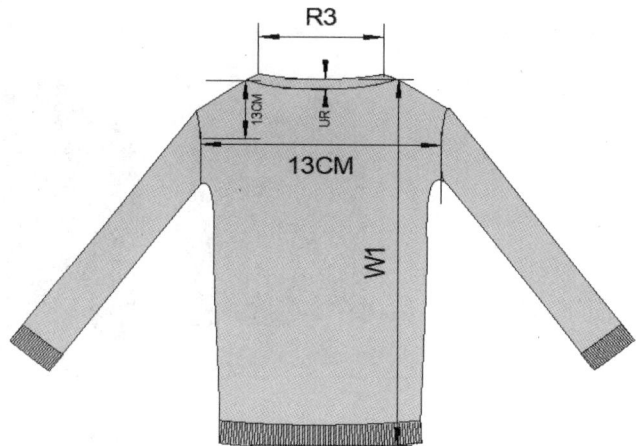

Uniquement pour le flechage de la prise de mesures
non conforme pour le style

**only for the points of measure
not corresponding for the style**

◇ Sample Ⅱ Technical instruction—specifications

FICHE DE MENSURATIONS / SIZE CHART SPECIFICATION

CREATED: 1/8/... UPDATED:
IPC: 23716
TECHNOLOGIST: C.DEBOSSCHERE
UR reference code: 23716

FOURNISSEUR / SUPPLIER: SPRINGWAY
DESIGNATION / DESCRIPTION: pull fantaisie

REFERENCE — France REF: — UK REF: — Belgique REF: — Saison/Season:

GRADING RULES — Jump grade at 18 >|< 20

			XS	S	M	L	XL	XXL	XXXL
	AH14							XXL	
	WW14								XXXL
TAILLE / SIZE Belgique	DUAL		34/36	38/40	42/44	46/48	50/52	54/56	58/60
TAILLE / SIZE France	DUAL		8	10/12	14/16	18/20	22/24	26/28	30/32
TAILLE / SIZE UK	DUAL		8	10	12				

Sketch Key	Description	Tolerance	XS	S	M	L	XL	XXL	XXXL	ARCHITECTURE [S/D/MH]	Grading below jump	Grading above jump	Offset at Jump Grade
A	½ tour de poitrine à la pointe emmanchure / Half chest width at armhole	± 1.0	42	46	50	55	61	67	73	D	2.00	3.00	0.00
D	½ ampleur base (haut) / Half hem width (Tops)	± 1.0	31	35	39	44	50	56	62	D	2.00	3.00	0.00
R	ecart encolure dos couture à couture / Back neck width seam to seam	± 0.5	21.5	22.5	23.5	24.5	25.5	26.5	27.5	D	0.50	0.50	0.00
S	profondeur encolure devant / Front neck drop	± 0.3	11.5	12	12.5	13	13.5	14	14.5		0.25	0.25	0.00
G	largeur épaule sous encolure / Shoulder length exc rib	± 0.3	7.5	8	8.5	9.25	10.25	11.25	12.25		0.25	0.50	0.00
kk	longueur manche (maille) / Sleeve length (knitted)	± 0.5	56.5	57	57.5	58	58.5	59	59.5		0.25	0.25	0.00
L	½ largeur haut de manches / Half Bicep	± 0.5	15	16	17	18.25	19.75	21.25	22.75		0.50	0.75	0.00
T	½ largeur emmanchure / Half armhole width	± 0.5	20	21	22	23.25	24.75	26.25	27.75		0.50	0.75	0.00
M	½ largeur bas de manche longue / Half cuff width long sleeve	± 0.5	7.5	8	8.5	9	9.5	10	10.5		0.25	0.25	0.00
E	carrure dos / Across back	± 0.5	34	36	38	40.5	43.5	46.5	49.5		1.00	1.50	0.00
W1	hauteur dos à la pointe épaule inférieur ou égal à 65cm / Length from SNP under or equal 65cm	± 1.0	59	61	63	65	67	69	71		1.00	1.00	0.00
TA	hauteur bord côtes poignet / Cuff rib depth	± 0.5	6	6	6	6	6	6	6		0.00	0.00	0.00
O2	hauteur bord cotes base / Welt depth	± 0.5	6	6	6	6	6	6	6		0.00	0.00	0.00
UR	hauteur de col milieu dos / CB collar depth	± 0.3	1	1	1	1	1	1	1		0.00	0.00	0.00

◇ Sample Ⅲ　Technical instruction—labelling

FOURNISSEUR / SUPPLIER	DESIGNATION / DESCRIPTION	REFERENCE	
SPRINGWAY	**pull fantaisie**	France REF:	**23716**
		UK REF:	
		Belgique REF:	
		Saison/Season: **AH14**	**WW14**

Care label

FRONT — 32mm

F-D-CH	BNL	UK
38/40	S	10/12

70mm maximum

Brand Label

Type: Refer to intern. labelling manual or Specific instruuctions given by the international buyer

Position: Refer to intern. Labelling manual

Care Label

Type: 100% Polyester, double faced with durable printing to 60°C washing (and dry cleaning where applicable).
[One side printed loop is also acceptable].

Artwork: As shown. Label length to be reduced where possible by removing vertical white-space. (minimum 32mm long).

Position: Refer to manual

Additional requirements

Spare buttons: to be sewn to care label (without covering any information) or separate label

Accessories: Spare sequins / beads / thread in small polybag kiballed to care label

BACK

Size architecture (Quantities on purchase order)

F-D-CH	BNL	UK			
34/36	XS	8	-	-	-
			-	-	-

F-D-CH	BNL	UK			
38/40	S	10/12	-	-	-
			-	-	-

F-D-CH	BNL	UK			
42/44	M	14/16	-	-	-
			-	-	-

F-D-CH	BNL	UK			
46/48	L	18/20	-	-	-
			-	-	-

F-D-CH	BNL	UK			
50/52	XL	22/24	-	-	-
			-	-	-

F-D-CH	BNL	UK			
54/56	XXL	26/28	-	-	-
			-	-	-

F-D-CH	BNL	UK			
58/60	XXXL	30/32	-	-	-
			-	-	-

UR reference Number / UR supplier code (please refer to the country purchase order) ⟶ 23716 /YYYYY

2 letter of the maker country code added in this place refer to the manual ⟶ ZZ

ABC Oral Practice

Forming a group of two or three partners. Try to work out a dialogue related to the theme of this unit and then perform it in class.

Self-Assessment

Evaluate your practice by marking the following each corresponding item and work out the total scores.

考核情境(分值)	考核要求(分值)	得分
语音语调(20 分)	发音准确，声音清晰(10 分)	
	语调自然，语速自然流畅(10 分)	
内容(30 分)	内容完整(10 分)	
	内容表述符合给定情境(10 分)	
	前后内容逻辑一致，内容之间无矛盾(10 分)	
语言描述(30 分)	用词规范、准确，无语法错误(20 分)	
	表述无歧义(10 分)	
其他(20 分)	无雷同状况(10 分)	
	无严重偏题(10 分)	
合 计(100 分)		

Task 4　Negotiation on Price

知识目标

◆ 学习国际贸易中价格谈判的内容
◆ 学习国际贸易中价格谈判的艺术
◆ 学习相关商务沟通的口头英文表达方式
◆ 学习相关商务沟通的书面英文表达技巧

能力目标

◆ 能够准备价格谈判材料
◆ 能够顺利开展价格谈判活动
◆ 能够使用双语顺畅地进行商务口头沟通
◆ 能够处理价格谈判往来信函

Basic Knowledge Review

1. What does negotiation on price involve?

In order to price your product properly and give accurate quotations, you need to determine your costs using three key elements: choosing the terms of the sale, selecting the payment method, and calculating taxes and tariffs.

To optimize your export pricing strategy, understand the total or "landed" cost of your export shipment to a foreign buyer. Buyers often ask for this information when negotiating a purchase and you can provide it on a proforma invoice.

The landed cost is the total price of a product once it has arrived at the buyer's doorstep. This includes: the original price of the product, insurance, freight, tariffs and taxes, and other fees.

In order to determine landed cost, both sellers and buyers should discuss the importing country's tariffs and taxes.

Here are some steps to follow.

(1) Find the importing country's Harmonized System (or HS) code for your product.

(2) Then, look up the tariffs and taxes for the country to which you are selling. Choose the best tariff rate for which your product qualifies.

Tariffs are charges imposed by countries on imported goods. They are organized by HS classification codes and often expressed as a percentage of the product's value. For example, ball point pens have a 5% tariff rate.

A duty is the actual amount of money owed due to the tariff. For example, the duty is $100.

The most common tariff categories are:
✓ General Rate—a standard tariff rate applied to countries that are not members of a World Trade Organization (WTO) or other agreement.
✓ Most Favored Nations or MFN Rate—a tariff rate for WTO signatory countries. This rate may change at any time within certain limits.
✓ United States of America Rate—this tariff rate is agreed upon in a U.S. trade agreement with another country or group of countries to be more advantageous than the MFN rate. It could be as low as 0%, or lowered over time to 0%, if your product meets certain requirements.

(3) Determine if other taxes and fees might be imposed. These can include customs fees, value added taxes (or VATs), or national/local sales taxes.

(4) Know if you or the buyer are obligated to pay the duties and taxes. This is often determined by the Incoterm used in the sales contract. You can also discuss payment options with your customs broker or freight forwarder.

When your product reaches a foreign market, the foreign customs authority will assess duties. Taxes and other charges are calculated separately.

2. How do we achieve the goal of negotiation on price successfully?
As pricing is often the most sensitive issue in business negotiations the subject should usually be postponed until all of the other aspects of the transactions have been discussed and agreed upon. To meet price objections, some exporters artificially inflate their price quotations. This enables them to give price concessions in the opening of the negotiations without taking any financial risks. The danger of this approach is that it immediately directs the discussions in to pricing issues at the expense of the other important components of the marketing mix. Generally such initial price concessions are followed by more demands from buyers that will further reduce the profitability of the export transaction.

(1) For instance, the buyer may press for concessions on :

✓   Quality discounts.

✓   Discount for repeat orders.

✓   Improved packaging and labeling (for the same price).

✓   Tighter delivery deadlines that may increase production and transport cost.

✓   Free promotional materials in the language of the import market.

✓   Market exclusivity.

✓   A long term agency payment.

✓   Higher commission rates.

✓   Better credit and payment terms.

To avoid being confronted by such costly demands, an exporter should try to determine the buyer's real interest in the product from the outset. This can be ascertained through appropriate questions but also must be based on research and other preparations before the negotiations.

After covering all of the non-price issue, the exporter can shift the discussions in the final phase of the talks to financial matches having a bearing on the price quotation. This is the time to come to an agreement on issues such as credit terms, payment schedules, currencies of payment, insurance, commission rates, warehousing costs, costs of replacing damaged goods and so on. Agreement reached on these points constitutes the price package.

(2) Hints for perfect price list :

Your price list is probably the single most important sale tool you have. If properly developed and presented it will do much to create a favorable impression, minimize costly errors and generate repeat business. Remember the following points while making your price list:

✓   Submit a typewritten list, printed on a regular bond paper and laid out simply and clearly.

✓   Prominently indicate the name of your company, its full address, telephone, fax numbers, e- mail address including the country and city codes.

✓   Fully describe the items being quoted and delivery schedule.

✓   Group the items logically.

✓   Specify whether shipped by sea or by air, FOB or CIF and to what port.

✓   Quote exact amount and quantity and not rounded-off figures.

3. What is FOB in International Commercial Terms?

Free on Board (named port of shipment): which is more commonly called FOB. FOB is the most common agreement between an international buyer and seller when shipping cargo via sea. This incoterm only applies to sea and inland waterway shipments. It's a popular way to arrange import of consumer goods from Asia to Europe. If you're finding a supplier through Alibaba, you might even see that there are separate FOB unit prices shown, as Alibaba recommends FOB as the best incoterm to use in many circumstances for example. Under FOB terms, the seller is effectively responsible for costs up to the point that the goods are loaded onto a ship, at a named port in the country of origin, and ready for transportation.

4. What are the Seller's Responsibilities under FOB terms?

FOB means Freight On Board or Free On Board. If terms of delivery of a transaction is on FOB means, the cost of movement of goods on board of ship is borne by the seller. Rest of all expenses to arrive the goods at buyer's premise has to be met by the buyer.

When entering into a sales contract with the incoterms of FOB, the seller assumes the following responsibilities:

(1) Export Packaging: arranging all export packaging so the cargo can be shipped safely.

(2) Loading Charges: any expenses incurred during the loading process at the seller's warehouse.

(3) Delivery to Port: trucking fees incurred while moving the cargo from the warehouse to the port of loading.

(4) Export Duty, Taxes & Customs Clearance: ensuring the cargo is exported correctly.

(5) Origin Terminal Handling charges: The seller is responsible for OTHC.

(6) Loading on Carriage: the seller is responsible for the costs incurred to load the cargo onto the carriage.

FOB - INCOTERMS® 2020

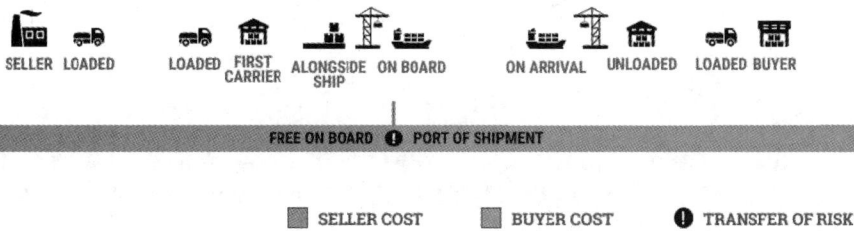

SELLER LOADED LOADED FIRST CARRIER ALONGSIDE SHIP ON BOARD ON ARRIVAL UNLOADED LOADED BUYER

FOB FREE ON BOARD ❶ PORT OF SHIPMENT

⬛ SELLER COST ⬛ BUYER COST ❶ TRANSFER OF RISK

5. How many parts does clothing industry FOB price include?

Generally speaking, clothing industry FOB price includes Material Cost, CMT Cost(Cost of Manufacturer), ACCESSORIES Cost, Transportation Fee, Custom Clearance Fee, Duty, Currency Rate, Test fee and Inspection Fee and Profit.

6. What is DDP and the composition of DDP costing?

DDP (named port of destination): Delivered duty paid (DDP) is a delivery agreement whereby the seller assumes all of the responsibility, risk, and costs associated with transporting goods until the buyer receives or transfers them at the destination port.

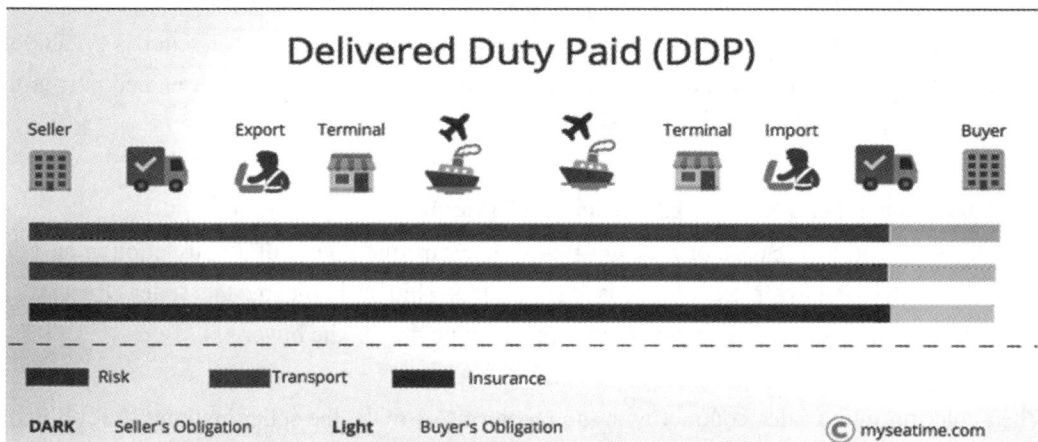

Generally speaking, DDP Price includes Materials Cost, CMT, Profit, Local Custom Clearance Fee, Shipping Fee, Insurance Fee, Destination Custom Clearance Fee and Local Transportation Fee.

7. What is CIF and the composition of CIF costing?

CIF(named port of destination): CIF is COST INSURANCE AND FREIGHT. Under CIF- terms, Seller must pay the costs and freight includes insurance to bring the goods to the port of destination. However, risk is transferred to the buyer once the goods are loaded on the ship.

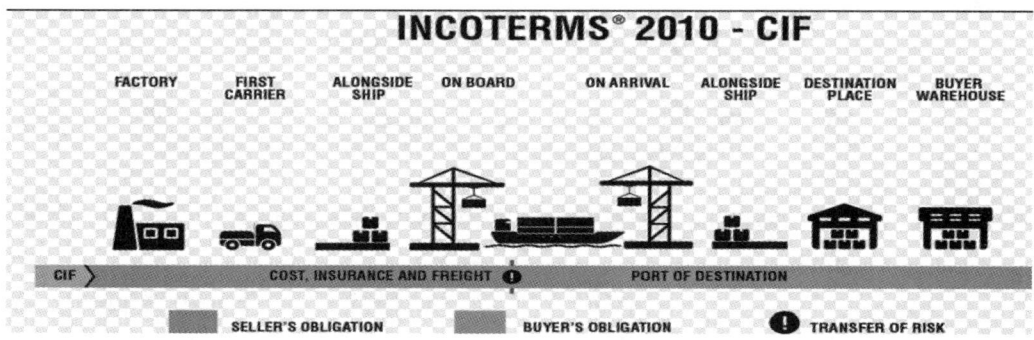

Generally speaking, CIF Price includes Materials, CMT, Profit, Local Custom Clearance Fee, Shipping Fee and Insurance Fee.

✍ **Task Descriptions**

美国斯特朗服装有限公司采购部经理 Betty 将在杭州红粉服饰有限公司的考察情况汇报给总部后，公司决定就某些产品询盘。

接下来的对话就是杭州红粉服饰有限公司代表与美国斯特朗服装有限公司代表围绕产品价格谈判等一系列活动展开的。

Negotiation on Price

Typical Dialogues

◇ Dialogue Ⅰ

这段对话的内容是：杭州红粉服饰有限公司外贸业务员 Dicky 与美国斯特朗服装有限公司采购部经理 Betty 谈论产品的质量问题。

A：杭州红粉服饰有限公司外贸业务员 Dicky

B：美国斯特朗服装有限公司采购部经理 Betty

A：Hello! Are you satisfied with this summer dress? This is our limited edition this year.

A：你好！对我们这款夏装还满意吧？这是我们今年的限量版呢。

B：Mm-hmm. Actually, this style is quite popular this year.

B：嗯嗯，其实这个款式今年还挺流行的。

A：Please believe that our quality is far superior to any other in its class.

A：请相信我们的品质远比同类其他产品优越。

B：Well, I'm sure your quality is absolutely out of question, but I think the price is still a little high.

B：嗯，相信你们的品质肯定没有问题，就是觉得价格还是有点偏高。

A：This dress is made of the most skin friendly fabric, and its dyeing is adopting the latest printing and dyeing technology. Look, this green is very pure, and will never fade after washing many times. Comparing with other goods of similar quality, I'm sure you'll find that our prices are reasonable and very competitive.

A：我们选用的是最亲肤的面料，染色也是采用了最新引进的印染技术，您看这个绿色很纯而且多次洗涤后都绝对不会出现褪色的问题。对比其它同质量的商品，我相信您一定会发现我们的价格是合理的，也是非常具有竞争力的。

B：Let me think about it for a while.

B：我再考虑一下吧。

◇ Dialogue Ⅱ

这段对话的内容是：杭州红粉服饰有限公司外贸业务员 Dicky 与美国斯特朗服装有限公司采购部经理 Betty 谈论订单的数量问题。

A：杭州红粉服饰有限公司外贸业务员 Dicky

B：美国斯特朗服装有限公司采购部经理 Betty

B：We are satisfied with this summer dress wear, but I don't think we can accept the price that high.

B：我们对这款夏装很满意，但不能接受这么高的价格。

A：Well, actually we offer quantity discounts. For instance, if your orders are over 10,000 pieces, you can receive a five percent discount.

A：嗯，实际上我们是提供数量折扣的。例如，如果您的订单超过 10,000 件，您就可以享受百分之五的折扣。

B：Five percent discount? It's still too high. Can't you make it cheaper?

B：九点五折？还是很高。你不能再便宜一点吗？

A：Sure. In addition, we'll further reduce the price for orders in large lots. So, you see, the more you order, the more you could save.

A：当然。我们还会针对大批量的订单再降低价格。所以，您看，您买得越多，省得就越多。

A：Well. Our company also has a special profit giving activity for old customers this time. If you place an order today, you can enjoy a 10% discount.

A：而且公司这次对老客户还专门有一个让利的活动。如果您今天下单，可以享受一个九折的优惠。

B：Mmm… Let me see if we need that many.

B：嗯……我看看我们是否需要那么多。

A：OK, just take your time, please sit down and have a cup of coffee.

A：好的，慢慢来，请坐下来先喝杯咖啡。

B：OK, thank you.

B：好的，谢谢。

◇ Dialogue Ⅲ

这段对话的内容：是杭州红粉服饰有限公司外贸业务员 Dicky 向美国斯特朗服装有限公司采购部经理 Betty 报 CIF 价。

A：杭州红粉服饰有限公司外贸业务员 Dicky

B：美国斯特朗服装有限公司采购部经理 Betty

A：Suppose we get down to business now.

A：正式会谈现在开始。

B：Yes, I have decided to increase the sale of your dresses at our end. Could you make an offer on CIF basis?

B：好的，我决定在我方市场增加你们裙子的销量。你们能报 CIF 价吗？

A：Our CIF price is 20 dollars per unit?

A：CIF 价是一件 20 美元。

B：The price you mentioned wouldn't be worthwhile for my company.

B：你们的报价对我公司来讲不合算。

A：This is a fair market price. What leads you think that we have to reduce our price?

A：这是市场价。你觉得我们为什么要减价？

B：Suppliers should try their best to reduce their CIF price, even make it lower than their home price.

B：供货商总是尽可能降低 CIF 价，甚至把价压得比国内售价还低。

A：But we give you quotations on the same basis as we quote in the domestic market.

A：但是我们的报价是和国内市场一样的。

B：That is the point. Your overhead is in your domestic price, but it can't be carried in the CIF price.

B：问题就在这，管理费算在国内价格内，不能放在 CIF 价内。

A：That's all right. I will look into it.

A：好的。我会考虑的。

B：I wish you would, because there is plenty of other business offering there.

B：但愿如此，因为我们还有其他的选择。

◇ Dialogue Ⅳ

这段对话的内容是：杭州红粉服饰有限公司外贸业务员 Dicky 与美国斯特朗服装有限公司采购部经理 Betty 讨价还价。

A：杭州红粉服饰有限公司外贸业务员 Dicky

B：美国斯特朗服装有限公司采购部经理 Betty

B：The price you quoted this year is rather on the high side.

B：你们今年的报价确实有点偏高了呢。

A：We're sorry, but the actual profit does not increase much.

A：真不好意思，其实实际的收益并没增加多少。

B：I wouldn't call this slight. The 20% increase is very troublesome to me. We need to talk it over some more.

B：可我真不认为这只是小幅度的涨价。这 20%的涨幅比较麻烦，我们需要再考虑一下。

A：OK. The cotton harvest is not so much this year, and the price of cotton fabric has increased a lot. Therefore, we have to raise the price.

A：好的。主要是因为今年的棉花收成是小年，棉质面料的价格提高了好多，我们不得不提价。

B：But your offer is much higher than your competitor's. Isn't there any room for further negotiation?

B：可是你们的价格在同行中都属于较高的了，难道没有再协商的余地了吗？

A：Sorry, this is our best price. We have introduced the top designers in Asia this year, thus our clothes are the best in both style and texture.

A：很抱歉啊，这已经是我们的最低价了。我们今年引进了亚洲最顶尖的设计师，请相信我们的服装无论在款式还是质地上都是最棒的。

B：How about this? You take 5% off the original offer, and I add 5% on the counter-offer.

B：在这样好不好？你们在原报价上减少 5%，我们则在原还价上增加 5%。

A：I'll check back with my head office, and I'll let you know as soon as I get an answer.

A：我回总部再和大家商量一下，会尽快给您消息。

B：Fine.

B：好的。

◇ **Dialogue　V**

　　这段对话发生在杭州红粉服饰有限公司外贸业务员 Dicky 与美国斯特朗服装有限公司采购部经理 Betty 之间，双方讨论实盘价格问题。

A：美国斯特朗服装有限公司采购部经理 Betty

B：杭州红粉服饰有限公司外贸业务员 Dicky

A：I have come about your offer on September 2.

A：我已经在 9 月 2 日收到你的报价了。

B：Oh, yes. We have the offer ready for you now. Here it is, 1500 pieces at 40 US dollars per piece, CIF San Francisco, for shipment during December, 2020. Other terms and conditions remain the same as usual. The offer is valid for three days.

B：哦，是的。我们现在开始报价：1500 件，每件 40 美元，CIF 旧金山，2020 年 12 月发货。其他条款和往常一样。该报价有效期三天。

A：Why, your price has gone up sharply! It is 20% higher than last year's. This is incredible!

A：啊，价格涨了这么多！比去年同期高出了 20%，这太不可思议了！

B：I am surprised to hear you say it. As you well know, there has been a strong demand for this kind of dresses and such a demand will certainly lead to increased price. Our price is more competitive than quotations you can get elsewhere.

B：听到你这么说，我感到很惊讶。如您所知，这款裙子很走俏，价格上涨是必然的。我们的价格相比其他地方还是很有竞争力的。

A：I don't think so. I must point out that some of the quotations we have received from other sources are lower than yours.

A：我不这么认为。我必须指出，其他厂家报价比你的要低。

B：But you must take the quality into consideration. To be frank with you, if we were not friends, we would hardly be willing to make you a firm offer at this price.

B：但你必须考虑到质量。坦白地说，如果我们不是朋友，我们是不会报这个实价的。

A：You are right. Yours is of high quality and we are old friends. But it will be very hard for us to push sales at this price. Perhaps I will have to try, I suppose.

A：这倒是的。质量不错，加上是老朋友。但以这个价很难卖得动。不过，我觉得还是要试一下。

B：Good. I am sure it is a wise decision.

B：我相信这是一个明智的决定。

👫 Roles Simulation

Suppose you are one of them in the following conversation. Try to read aloud and practice the underlined sentences.

　　这段对话发生在杭州红粉服饰有限公司外贸业务员 Dicky 与美国斯特朗服装有限公司采购部经理 Betty 之间。双方围绕商品价格问题展开一系列的讨论。

A：美国斯特朗服装有限公司采购部经理 Betty

B：杭州红粉服饰有限公司外贸业务员 Dicky

A：Your price is much higher than you quoted last time.

A：你方价格比上次报价高多了。

B：As you ask for special packing and delivery, our price should include the extra charges. You can't expect us to keep to the original price, can you?

B：由于你方要求特别包装和发运，我们的价格就把额外的费用包含在内了。你总不能还要求我们维持原价吧。

A：You must know that all companies are cutting their prices in order to get a large market share.

A：你要明白，所有公司都在减价，以便获得更多的市场份额。

B：That's right. But extra services deserve extra pay. With the special packing and delivery you require, I don't think any other company can offer such a competitive price as ours.

B：没错。可是，额外劳务就得有额外费用。你方要求特别包装和发运，我想哪个公司也不可能提供像我们这样有竞争性的价格吧。

A：To be honest, we won't be able to make a profit at your price. Can't we find a price that is good for both sides?

A：坦率地说，按照你方价格我们将无利可图。我们能否找到一个彼此都有利可图的价格呢？

B：We don't do much bargaining. We go in for business on the basis of mutual benefit.

B：我们不喜欢讨价还价。我们做生意从来讲究互利。

A：Sellers never admit their prices to be higher than their competitors. But I'm afraid this time if you don't bring your price into line with the prevailing market, you'll stand no chance.

A：卖主从来不会承认他们的价格比对手高。不过这一次如果你的价格不降到与市场一致，恐怕就没希望了。

B：You seem to be hinting at something. As our old customer and friend, you might very well speak out your mind in a more direct way.

B：好像你话中有话。作为老客户，你不妨直截了当，想说什么就说什么嘛？

A：By all means. In my opinion, you must reduce your price by 5% in order to stand competition from other suppliers.

A：那我就直说了。我认为要想经得起对手竞争，你方价格必须降低 5%。

B：5%? I'm afraid you are asking too much. To tell the truth, ours is the rock-bottom price.

B：5%？你太过分了吧。老实说，这是我们的底线了。

A：We have had a very good relationship over the past years. I think this transaction would be more promising if you could make an appropriate reduction.

A：过去几年我们双方合作愉快，如果这次您能适当降价，交易将大有前途。

B：OK. Considering our good cooperation, we are prepared to make some concessions.

B：好吧。鉴于你我良好的合作关系，我们愿意做些让步。

A：Then how much can you go down?

A：能降多少呢？

B：1% off the original price.

B：在原价基础上降 1%。

A：1%? <u>Your reduction is too modest.</u> Can you take one more step forward and give us 3%?

A：1%？<u>太少了吧。</u>能不能再多让点，给 3%？

B：3% won't leave us anything.

B：3%我们就一点利都没有了。

A：I think it unwise for us both to insist on our own views. How is this, then? <u>To pull this transaction through, let's meet halfway? You see, joint efforts would help carry us each one more step forward.</u>

A：我认为双方都固执己见是很不明智的。你看这样行不行？<u>为了达成交易，双方再各让一半怎么样？你瞧，共同努力会使得我们彼此向前迈进一步。</u>

B：You certainly have a way of talking me into it. All right, I give up. Let's meet each other half way. <u>Our maximum is 2%. We can't go any further.</u>

B：你这人说服别人还真有一套。得，我放弃。咱们各让一半。<u>最多 2%，不能再少了。</u>

A：All right. <u>As a token of our sincerity, we accept the price.</u>

A：行。<u>为表示我们的诚意，我们接受了。</u>

💝 *Typical Expressions*

1. I enclose details of our products. 附件是产品的详细信息。

　Please confirm the information. 请确认信息。

　Thank you for your inquiry. 谢谢您的询盘。

　We will be delighted to /It will be a pleasure to receive your inquiry. 很高兴能收到你的询盘。

　I regret that I could not accept the quotation. 很抱歉不能接受报价。

2. Our products are all quality goods and well thought of abroad. 我们的产品全都是优等品，在海外市场声誉不错。

3. If we order 1000 pieces or more, how much can you bring the price down? 如果订购 1000 件以上，价钱可以降多少？

4. If you order in large quantities, I think a discount would be possible. 若是惠予大量订购，敝公司当可酌减价格。

5. Your price is too much on the high side. Would you please cut the price down to $ 15 per piece? 你们的价格偏高太多。可否降到每件 15 美元？

6. I'll respond to your counter-offer by reducing our price by three dollars. 我减价 3 美元以回应你们的还价。

7. I mean both parties should compromise a little and meet each other half way. 我的意思是双方稍加妥协，各让一半。

8. How about this? You take 5% off the original offer, and I add 5% on the counter offer. 这样好不好？你在原报价上减少 5%，我则在原还价上增加 5%。

9. I'm afraid that there is no room to negotiate the price. 很抱歉，这个价格没有再商量的余地了。

10. We have to raise the price because the cost of raw materials rose. 由于原料价格上涨，我们不得不提价。

Foreign Trade Letter

◇ **Sample　Saying "No" letter**

How and when you have to say "NO".

1. If buyers keep changing, and it is their responsibility, please request the reasonable delivery date.

2. If your buyers increase the cost, please re-quote the price to them with the reason why you increase the price.

发件人：XXXXXX
Subject: Re: 进程
To: XXXXXX
Cc: XXXXXX
Dear XXXXXX,
How are you?

I am not happy about this program, for this kind of order, we can not accept changing the measurements and gauge back and forth, there are a lot of work, these are not easy styles, each time, we need remake the knit plan, knitting plan and knitting to finishing.

For delivery, we cannot keep the previous one, we need 45～60 days to make the production after you confirm the pps.

Highly appreciated for you kindly understanding.

Best regards!
XXXXXX

■ **Notes：**

1. knit plan: 针织平面图

2. pps: pre-production sample 产前样

Foreign Trade Documents

◇ Sample Ⅰ　FOB costing sheet

ZONGSEN TEXTILE CO.,LTD				
COSTING SHEET(CMT ONLY)				
STYLE NO.:	15PE015A		DATE:	11/28/2014
FABRIC COMPOSITION				
FABRIC WIDTH: 1.5ms				
COSTING ITEM	FABRIC CONSUMPTION(METRE)	PRICE/M	TTL	PHOTO
FABRIC				
CMT			￥19.00	
FINISHING			￥4.00	
BUTTON			￥1.00	
ACCESSORY			￥3.00	
CUSTOM FEE			￥2.00	
TRANSPORTATION FEE			￥2.00	
PROFIT			￥6.00	
TOTAL			￥37.00	
FOB Guangzhou US$			USD$6.17	Fabric Supplier Information
REMARK:				
1) The price base on the sample is the middle sample of the order.				
2) The price based on order quantity 1000pcs or above.				
3) Fabric minimum is 1000ms per color if no stock.				
4) VAT is free.				

■ Notes：

1. costing sheet(成本核算表)：成本核算表是生产商或工厂计算产品使用的物料或订单用料的所有工序中产生的 overhead(费用)，即生产成本。这里的生产成本是既包括用于加工产品的物料等直接成本，也包括保障生产产品顺利进行的间接成本，比如行政管理费用、交通运输费用、关税、利润、人工费用等非生产性的成本。间接费间虽然不直接参加产品的生产过程，也要以一定比例计入成本核算中。间接成本通常以直接成本的 10～15% 计算。行业不同，产品的成本要素也不相同，自然成本核算构成及各部分占比也不尽相同。

2. COSTING ITEM：成本要素

3. FABRIC：布料

4. FABRIC COMPOSITION：布料成分

5. CMT：cutting, making and trimming 衣服裁剪加工费

6. FINISHING：后处理费用

7. BUTTON：纽扣

8. TRANSPORTATION FEE：运输费用，这里指本订单报价的工厂到装运港。

9. ACCESSORY：辅料

10. CUSTOM FEE：报关费

11. FABRIC WIDTH：布封，指布的宽度。

12. fm：from

13. pcs：pieces 件

14. Fabric minimum is 1000ms per color if no stock. 如无存布，布料的起订量每种颜色不少于 1000 米。

15. VAT is free：免收增值税。VAT：Value Added Tax 增值税

◇ Sample Ⅱ　FOB price quotation sheet template

东莞市智恒针织制衣有限公司

地址：东莞市寮步镇凫山村长富工业区兴新街 2 号 3 楼

电话：0769-81169853　传真：0769-81239727

毛衣报价表

日　期：	2021/12/30	针种：	12　针	
客　户：	SPRINGWAY	重量：	167　克	
款　号：	SU-H19-PU16-LIF/SU	毛料供应商信息：		
毛料成分&价格：	A:2/32 40%粘胶　25%尼龙 18%羊毛　17%腈纶	A: 72 元/kg		
	B: 金银线	B: 48 元/kg		
	C:	C:		
成衣毛料 比例&毛线	A: 毛料 15 元/件			
	B: 金银线 0.3 元/件			
	C:	辅料供应商信息备注栏		
花边价格：				
吊染：				
拉链价格：				
基本辅料：	2.5 元			
其他辅料：				
测试费用：				
运费及报关费用：				
工厂报 FOB 价：	56.4 元			
工厂报 CMT 价：		织机：9.6 缝挑：8 后整：8 利润：12		
智恒利润：		备注：　　　　车挂衣带：1 元		
FOB 价格(RMB)：				
DOP 价格(USD)：		汇率：		
最终与客人确认价格：		确认人及日期：		
外　发　价　格				
外发 FOB 价格：		外发 CMT 价格：		
		CMT 价格 = 车工价 + 辅料费用		

审批人：　　　　　　　　　　　　制表人：

日　期：　　　　　　　　　　　　日　期：

Oral Practice

Forming a group of two or three partners. Try to work out a dialogue related to the theme of this unit and then perform it in class.

Self-Assessment

Evaluate your practice by marking the following each corresponding item and work out the **total scores.**

考核情境(分值)	考核要求(分值)	得分
语音语调(20 分)	发音准确，声音清晰(10 分)	
	语调自然，语速自然流畅(10 分)	
内容(30 分)	内容完整(10 分)	
	内容表述符合给定情境(10 分)	
	前后内容逻辑一致，内容之间无矛盾(10 分)	
语言描述(30 分)	用词规范、准确，无语法错误(20 分)	
	表述无歧义(10 分)	
其他(20 分)	无雷同状况(10 分)	
	无严重偏题(10 分)	
合　　计(100 分)		

Task 5 Negotiation on Shipment

知识目标

◆ 学习国际贸易中运输谈判的内容
◆ 学习国际贸易中运输谈判的艺术
◆ 学习相关商务沟通的口头英文表达方式
◆ 学习相关商务沟通的书面英文表达技巧

能力目标

◆ 能够准备运输谈判材料
◆ 能够顺利开展运输谈判活动
◆ 能够使用双语顺畅地进行商务口头沟通
◆ 能够处理运输谈判往来信函

Basic Knowledge Review

1. What does negotiation on shipment involve?

Negotiation on shipment is usually related to shipping terms. Shipping terms are also called INCOTERMS.

Incoterm is the elided word that shortens International Commercial Terms. They are 3 letter abbreviations recognized throughout the world. They tell each party concisely what is expected of them in selling and in contract negotiations.

In all international transactions, shipping can be paid for and done by either the shipper or the consignee. For example, the shipper can pay for everything: from their dock to the consignee's dock with customs clearance and duties paid on their behalf. This is called delivered Duty Paid (DDP). Or the shipper can do none of the shipping. The buyer picks up the freight from the shipper's factory dock. This is called ex-works (EXW). In between these two extremes, there are a number of places where the shipping can pass from one party to another. The scope can end once loaded on the carrier (FOB) or at a named place (DAT) or at a port or terminal (CFR and CPT). There are about a dozen of these shipping terms.

Basically, incoterms indicate three things.

✓ Who arranges for transport and the carrier

✓   Who pays for the transport

✓   Where/when does title (ownership) of goods transfer from seller to buyer

Consequently, shipping terms tell where costs are transferred and where the risk is transferred from the shipper to the consignee. Therefore, shipping terms are NOT optional.

Even when they are not stipulated or mentioned, both parties have expectations about the shipping. Ignoring the shipping terms only invites confusion at best and legal trouble at worst. Every sale, every quote, and every international contract here goods are exchanged must have incoterms.

2. How do we achieve the goal of negotiation on shipment successfully?

Global trade has thrived on the back of shipping and logistics companies for hundreds of years. As a result, the shipping industry has developed a unique culture and language. Here are some tips to clear the obstacles of negotiation on shipment.

(1) Learn shipping terms

If you don't understand the industry's regular terms, you risk making a deal that can be negative for your business. Terms such as CIF (Cost, Insurance, and Freight) and FOB (Free on Board) are used freely in the shipping industry.

(2) Research Possible Hidden Costs

One of the things that can diminish your profits is hidden logistics costs. The final price for your products should factor in all the costs you expect to incur before delivery. Talking to the different authorities that can come into contact with your products in transit can clarify your overall costs. Knowing the factors affecting your shipping rates can help you negotiate to reduce hidden costs.

By working closely with your shipping company, you can identify smart ways to cut down your costs. Many companies have different shipping rates based on weight as well as box dimensions. If you use the standard boxes the shipping company provides, you can save a few dollars on each load.

(3) Use Third Party Logistics Providers (3PL)

Third Party Logistics Providers (3PL) provide a useful service by shipping your cargo via their own intermodal networks. A trusted 3PL provider can save you time and money while allowing you to focus on your core business.

3PLs allow you to negotiate with one service provider who can manage all the regulatory and intermodal networking issues you may face. Further, your 3PL can help you connect and share shipping costs with other dealers in your vicinity.

Task Descriptions

杭州红粉服饰有限公司代表与美国斯特朗服装有限公司代表就价格谈判成功之后，开始进入产品运输相关问题的谈判。

接下来的对话就是杭州红粉服饰有限公司代表与美国斯特朗服装有限公司代表围绕产品运输谈判等一系列活动展开的。

Negotiation on Shipment

Typical Dialogues

◇ **Dialogue Ⅰ**

这段对话发生在杭州红粉服饰有限公司外贸业务员 Dicky 与美国斯特朗服装有限公司采购部经理 Betty 之间，双方讨论货物外包装问题。

A：美国斯特朗服装有限公司采购部经理 Betty

B：杭州红粉服饰有限公司外贸业务员 Dicky

A：I think your silk garment is superb. The traditional hand-embroidered design is an irresistible appeal to us Americans. But do you mind if I give you a little suggestion?

A：我觉得丝绸衣服不错，传统的手工刺绣很受美国人喜欢。介意给您提点建议吗？

B：Not at all. We appreciate any kind of suggestions or comments. They would be great help to our future work.

B：当然不介意。您的建议对我们帮助很大。

A：Well, there is no question of your garments. But have you ever thought of improving the packing of them? As far as I'm concerned, packing is as important as the products themselves. Without good and attractive packing, the buyers just ignore your products or even refuse to have a look at them even if they are of best quality. Who knows what kind of product is wrapped inside with such a poor packing?

A：衣服没问题，但是您想过改一下包装吗？包装和产品一样重要，没有好看的有吸引力的包装，买家可能连产品都不会看一眼。包装那么差的产品能包出什么好产品？

B：You sound really reasonable and convincing. Could you be more specific?

B：有道理。有什么具体建议吗？

A：I find you have packed your garments in plain simple plastic poly-bags. To make them more attractive, elegant, and expensive, I suggest you use double-packing, i.e., use an outer cardboard box with window display which can provide a look inside. The card-board box should be designed with exquisite style to match the excellent embroidery of the garment. In one word, the packing should be tasteful and eye-catching. Attractive packing promotes the sales.

A：我发现您把衣服简单地放在单层的塑料袋里。我建议改用双层包装，在外面加个有窗户的纸箱，卡盒设计精致些，这样看起来优雅、贵重更能吸引人。这样的包装才能配得上刺绣衣服。总之，好马配好鞍，包装促销售。

B：That's really a great idea.

B：好主意。

◇ Dialogue Ⅱ

这段对话发生在杭州红粉服饰有限公司外贸业务员 Dicky 与美国斯特朗服装有限公司采购部经理 Betty 之间，双方讨论提前发货的问题。

A：美国斯特朗服装有限公司采购部经理 Betty

B：杭州红粉服饰有限公司外贸业务员 Dicky

A：When can you ship the goods?

A：什么时候可以发货？

B：In December.

B：12 月。

A：Is there any possibility for you to ensure prompt shipment?

A：能保证及时发货吗？

B：I'm afraid not. Our manufacturers are fully committed at the moment. We dare not commit ourselves beyond what the production schedule can fulfill.

B：恐怕不行。制造商已全力组织生产了，我们不敢保证生产计划之外的生产任务。

A：That will be too late. You know, Christmas is coming, we want the goods on the market before the end of November to catch up with Christmas sales. Anyway, can you do anything to advance the shipment to November?

A：那就太晚了。你知道，圣诞节就要到了，我们希望这些商品在 11 月底之前赶上圣诞节的销售。不管怎样，你能把发货提前到 11 月吗？

B：The best we can do is to effect shipment at the end of November.

B：我们所能做的就是在 11 月底发货。

A：If you manage to deliver the goods two or three weeks earlier, everything will be better.

A：如果你能提前两到三周交货，事情会好很多。

B：I'll get in touch with the manufacturer and ask them to try their best to go advance the shipment to the middle of November.

B：我马上跟厂商联系，尽量把发货的时间提到 11 月中旬。

A：That's very kind of you. I'm looking for receiving your advice of shipment as early as possible.

A：太好了，期待尽快收到装船通知。

◇ Dialogue Ⅲ

这段对话发生在杭州红粉服饰有限公司外贸业务员 Dicky 与美国斯特朗服装有限公司采购部经理 Betty 之间，双方讨论发货时间的问题。

A：美国斯特朗服装有限公司采购部经理 Betty

B：杭州红粉服饰有限公司外贸业务员 Dicky

B：You will be very happy to hear that your order is ready for shipment now.

B：订单开始装货了。

A：That's good news.

A：真是个好消息。

B：Please tell me when and how you would like the goods delivered.

B：请告诉我们时间和交货方式。

A：We need the shipment in a hurry. Please send it in the fastest possible way. How fast do you think we could have this order?

A：我们需要尽快安排发货吧。你看最快我们多久能拿到货？

B：In five working days, if we ship by air.

B：空运的话，五个工作日吧。

A：Maybe we had better do that.

A：那就空运吧。

B：Of course we can. But according to the contract, you pay the extra freight and insurance if you want the goods to be sent by air. That might cost quite a bit more I am afraid.

B：当然可以。合同约定，如果空运，你方需要支付额外的运费和保险。这恐怕会增加成本。

A：That's OK. We would like to receive the shipment by May 1.

A：没关系。只要 5 月 1 号之前能收到货就行。

B：No problem! As to the freight, I'll check the rates and call you back.

B：没问题。核对好运费之后，马上回你电话。

◇ Dialogue Ⅳ

这段对话发生在杭州红粉服饰有限公司外贸业务员 Dicky 与美国斯特朗服装有限公司采购部经理 Betty 之间，双方确认发货的问题。

A：美国斯特朗服装有限公司采购部经理 Betty

B：杭州红粉服饰有限公司外贸业务员 Dicky

A：I'm calling to see what happened to our last order.

A：我打电话是想问上次订单的事。

B：Everything goes well.

B：一切顺利。

A：When do you think you could make a delivery?

A：那什么时候可以交货？

B：That is scheduled for shipment the day after tomorrow. Is this a rush order?

B：安排后天发货，很急吗？

A：Kind of. You know, we need it.

A：有点急。你知道，我们需要这批货。

B：It will be there. We'll get it right out.

B：我们会安排好的。

A：By the way, I'd like to remind you that we need our order number on the outside of each box. It makes it easier for us when we get the order on the dock.

A：顺便说一下。我想提醒你一下，把订单号贴在盒子外面。这样码头上拿货更方便。

B：That will be no problem. I'll take care of it for you.

B：没问题的。我会特别注意的。

A：Furthermore, Warn the trucking company to take it easy.

A：还有，告诉运输公司小心轻放。

B：I certainly will.

B：我会的。

◇ Dialogue Ⅴ

这段对话发生在杭州红粉服饰有限公司外贸业务员 Dicky 与美国斯特朗服装有限公司采购部经理 Betty 之间，双方确认运输的具体细节。

A：美国斯特朗服装有限公司采购部经理 Betty

B：杭州红粉服饰有限公司外贸业务员 Dicky

B：Now you see, this transaction is concluded on F.O.B terms. As the buyer, you are to charter a ship or book the shipping space.

B：您知道，这笔买卖是按装运港码头交货条件成交的。作为买方，你们应该租船或者订舱。

A：That's understood.

A：那是不用说的。

B：Then you should send us a Shipping Instruction so that we can make necessary arrangements for shipment beforehand.

B：然后你们还要给我们寄一份装船指令以便我们可以提前做好装船准备。

A：That's also understood. Will you make clear to me your responsibility?

A：那当然也不用说了。请您也清楚地向我们谈谈贵方的责任好吗？

B：We'll see that the goods pass over the ship's rail and bear all the charges up to the time the goods are on the hooks. Our responsibility ends there.

B：我们将负责将货物送过船舷，并承担货物上钩之前的一切费用。我方责任到此为止。

A：As I understand it, you are also to give us a Shipping Advice to inform us of all the detailed information related to the shipment of the goods. Is that so?

A：据我所了解，你们还应该寄给我们一份装船通知用以通知我方关于装运的详细情况。是吧？

B：Yes. We'll fax you immediately after the shipment.

B：是的。我们将在装船后立即按要求用传真将这些情况通知你方。

A：Excellent. So far we have a mutual understanding on everything concerning our respective responsibilities.

A：好极了。至今我们双方对于彼此责任的理解都是一致的。

🏃 Roles Simulation

Suppose you are one of them in the following conversation. Try to read aloud and practice the underlined sentences.

这段对话发生在杭州红粉服饰有限公司外贸业务员 Dicky 与美国斯特朗服装有限公司

采购部经理 Betty 之间，双方讨论运输细节。

A：美国斯特朗服装有限公司采购部经理 Betty

B：杭州红粉服饰有限公司外贸业务员 Dicky

A：I'm glad we have settled the questions of quality, quantity, packing and price. Today, let's talk about the shipment.

A：很高兴我们已经解决了品质、数量、包装以及价格方面的问题。今天，我们来谈谈装运事项吧。

B：Yes. How about defining the time of shipment first?

B：是啊。我们先把装运时间确定下来怎么样？

A：Good. When could you make a delivery?

A：好啊。你们什么时候可以交货呢？

B：Is this a rush order?

B：你们这批货要得很急吗？

A：Kind of. It would be easier for us if you would ship earlier.

A：有一点吧。如果能够早点出货的话我们会轻松一些。

B：We can deliver the goods in two months. That is in November.

B：我们可以在两个月后交货，也就是 11 月份。

A：That's too late. You know, November and December are our peak seasons.

A：那太晚了。你知道，11、12 月是我们的旺季。

B：That's exactly the time our goods reach your market.

B：那恰好是我们的货物抵达你方市场的时候。

A：Oh, you don't know it. Our Customs formalities are rather complicated. Besides, the flow through the marketing channels involved takes at least a couple of weeks.

A：噢，你不知道。我们的海关手续相当复杂。另外，相关营销渠道的流通，起码也要花上两个星期。

B：So?

B：那您的意思是？

A：The goods must be shipped before October，or they will not be in time for the selling season.

A：10 月份之前货必须装上船，否则我们就赶不上销售季节了。

B：But our factories are fully committed for the third quarter.

B：但是我们工厂第三季度的任务已经排满了。

A：You certainly realize that the time of delivery is a matter of great importance to us. Can you find some way to get round your producers for an early delivery?

A：你们一定可以意识到交货时间对我们来说真的是一件很重要的事情。你们能不能想办法让工厂早点交货呢？

B：By the middle of October I think. That's the best we can promise.

B：10 月中旬吧，我想。那是我们能够答应的最早的时间了。

A：Good. Let's call it a deal.

A：好！就这么定了！

Typical Expressions

1. We hope the goods will catch up with the season, can you ship them all at once? 我们希望这批货能赶上销售季节，你们能立即安排运输吗？

2. How are you going to handle the shipping on this? Do you want the cargo to go by sea or air? 这批货你们打算怎么运，海运还是空运？

3. There may be some quantity difference when loading the goods, but not more than 5%. To make it easier for us to get the goods ready for shipment, we hoped that partial shipment is allowed. 货物装船时可能会有一些数量出入，但不会超过 5%。为了方便我方备货装船，希望允许分批装运。

4. We'd like to designate Singapore as the loading port, for it has larger handling a full charter. 我们指定新加坡为装运港，因为它承租能力更强。

5. The goods will be transhipped in Hong Kong. 货物将在香港转运。

6. We have not received your shipping advice yet, please inform us whether shipment has been affected. 我们还没有收到你们的装船通知，请来信告知运输是否已经装运。

7. We will arrange the shipment as soon as we receive your L/C. 一收到你方信用证，我方将立即安排装船。

8. Would you please handle all the shipping formalities and insurance, and send us seven copies of the bill of lading, five copies of the commercial invoice, and the insurance certificate? 请及时办理装运和保险手续，并寄给我方 7 份提单、5 份商业发票和 1 份保险凭证行吗？

9. We are pleased to inform you that we have booked shipping space for our order No.9800 of dresses on SS Shan at Shanghai Port. 很高兴告知你方订单号 9800 的服饰已经订好了舱位：上海港货轮号"珊"。

10. As per terms of the contract, we will forward you by air mail a full set of non-negotiable documents immediately after the goods are loaded. 根据合同条款，货物装船后，我方将一整套不可转让的单据用航空邮件寄给你方。

Foreign Trade Letters

◇ Sample Ⅰ Apply for shipping

At 2021-05-23 15:50:52 "Simi" wrote:

Dear Julia,

We have below the goods to be shipped in ETD June 11 vessel total of 52.5CBM, we want to take a 40'container at Bollore. Please kindly check and help confirm back us, many thanks!

LIF/02/2020   8CBM
LIF/04/2020   4CBM
LIF/05/2020   10CBM
LIF/06/2020   2CBM
LIF/10/2020   7CBM
LIF/11/2020   2.3CBM
LIF/12/2020   1.5CBM
LIF/18/2020   0.6CBM
LIF/19/2020   0.7CBM
LIF/25/2020   5CBM
LIF/26/2020   3.4CBM
⋮
Total of 52.5 CBM
Best regards!
Simi

◇ Sample Ⅱ   Confirm shipper with buyer

Dear Simi,
Ok, please make booking with Bollore.
Thanks and best regards.
Julia

◇ Sample Ⅲ   Place booking to shipper

Dear Sally,
Attached please find the booking for ETD2021/6/11vessel, please kindly check and help release the SO to us asap, many thanks!
Best regards!
Simi
Attachment:

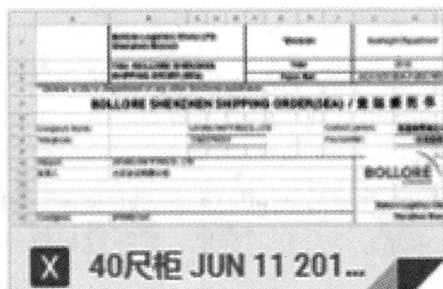

■ Notes：

SO: SHIPPING ORDER 装货单

◇ Sample Ⅳ　Apply for SO

Dear Sally,

麻烦尽快提供更新的 SO 给我们，谢谢！

另请问补料时间可否截至 6 月 6 日下午 16：00，因为现在还没收到 SO，我们的货也预计到明天下午可以全部装好柜，请帮忙，谢谢！

Best regards

Sasa

◇ Sample Ⅴ　SO information from shipper

Dear Customer:

Pls find the SO as attachment and confirm well received by return.如果要取消做柜赶不上船期，请至少在 SI CUT OFF 前三天通知我司，谢谢。

SO：XXXXX

V/V: APL SQ: XXXXXXXXX

CARRIER	SERVICE	Vessel	Voyage	SI CUT 12:00	CY CLS	YANTIAN	LE HAVRE
EMC/APL/ CMA/OOCL	FAL3	CMA CGM MARCO POLO	0kn9877LK	6/6	6/9	6/11	7/6

SI/VGM CUT: 请务必提供 6 位数字 HS CODE

CY Open(开仓)：6/3 (Do not pick up container before this date, if EXW pls instruct the pick up date) 请不要在开仓前提柜，EXW 条款需我司安排拖车请务必收到 SO 后立即回复提货时间和货物重量，否则车行旺季会耽误派车，谢谢；需我司报关清货当天必须尽早提供报关资料，不要耽误。

CY Cut Off(截关)：

ETD(预计开船日)：

ETA(预计到港日)：

Need you help to provide S/I before above deadline and please pay attention to below points:

1. Please double check the container type /POL/POD before arrange EIR, and question please let us know at once or any liability have to be bear by your good company. 请收到 SO 后仔细检查柜型、起运港和目的港，如果有任何问题，请立即通知我司修改；否则，因 SO 错误所导致的一切责任概由贵司承担。

2. Once confirm S/O is correct please arrange EIR at the first time otherwise may can't get equipment as space is very tight.请在确认收到的 S/O 信息正确后立即打单，以免因为爆舱或缺柜等原因造成无法打单，延误船期。

3. Please note CY is just open on xxxx so don't return container before this time and free detention is 3 days, it will cause surcharges at your account if over it. 开舱时间为 XXXX，请不要早于开舱日还柜。场外免柜租时间为 3 天，如果有超期柜租或舱租产生，将由贵司承担。

4. Please DO submit S/I before above S/I cut time or the container may can't be loaded on intended vessel, the related surcharges and responsibilities have to be bear by your good company. 请按时补料，否则柜子可能无法上船，由此产生的所有责任费用将由贵司承担。

5. 司机在还重纸上所写的毛重与补料时实际净重相差超过柜重±3 吨，我司将收取每个柜 RMB500 或以上的相关额外费用。

Bay plan weight 客户在还重柜时在重柜纸上所填写的毛重=货物重量+货柜重量

Manifest weight 客户在做提单补料时所提供的净重(货物重量)

Thanks and Best Regards!

◇ Sample Ⅵ   Submit SI and VGM to Shipper

Dear Sasa,

麻烦请立刻提供 SI and VGM。谢谢！！！

Holiday Notice

Our office will be closed from Jun 07th (Fri) to Jun 09th (Sun), total 3 days. for Dragon Boat Festival, and resume work on Jun 10th.

Best Wishes and Zhu Shi Shun Li!

通知：为了积极响应国家税务总局《关于推行通过增值税电子发票系统开具的增值税电子普通发票有关问题的公告》，我司决定用电子发票来替代原纸质增值税普通发票。如无特殊原因，我司将不再提供纸质增值税普通发票。特此通知，感谢您的配合！

Best Regards，

XXXXXXX

Seafreight Operation-Shenzhen

XXXX   Logistics China

Room 628, China XXXXX   Building, No 2002, XXX Middle Road, Shenzhen-CHINA

XXXX(上海)国际货运有限公司深圳分公司

深圳市福田区深南中路 XXXX 号 XXXX 大厦 7 楼 728 室

T(+86)XXXXXXX Ext. XXXX-Direct Line(+86)XXXXXXX

Sally XXXXXXX

■ Notes：

1. SI(shipping instruction)：提单补料，指订舱方根据提单要求向船运公司提供的相关货物的详细资料，比如柜号、封号、毛重、总立方数、唛头、货物描述等。货代根据这些资料制作出提单确认件，由出口商与 L/C 或合同核对无误后，给出正本提单和船证明。

2. VGM(Verified Gross Mass)：集装箱称重规定。核实集装箱总重量(VGM)的规定旨在减少由于集装箱总重的误报造成境内、外人员，内陆和海上货物以及设备存在风险并导致事故的发生。VGM 托运人通常是海运提单或者其他相关运输文件的托运人。VGM 称重方式有两种：一是全部货物装载完成之后，称量整个集装箱；二是集装箱内所有货物重量，再加上集装箱的皮重以及附加装载设备的重量。集装箱重量经码头工作人员核实后，可以登船。

◇ **Sample Ⅶ   Confirm container**

Sent：2019 年 6 月 5 日 16:32

To: XXXXX

Cc: XXXXX

Subject: 回复：RE：AJD0341592-232EBB

Dear XXXX,

已经可以了，谢谢！

箱　主：	CMA		打印日期: 2019-06-05
提箱地点:	盐田码头	25180000	
	1. 提箱地点有效期至2019-06-06 14:00 2. 逾期请按照以下备注要求在提柜前一工作日短信约柜		AJD0341592-232EBB

Best Regards

XXXX

◇ **Sample Ⅷ   Document deadline**

Dear XXX,

直接重新打单即可，谢谢！最迟明天下午三点提供一次性补料，谢谢！

Best Regards，

XXXX

Sea freight Operation-Shenzhen

■ **Notes：**

出口方如收到船运公司提示需补充材料的信息，应立即补料。如错过补料时间，出口方将会收到诸如"由于贵公司已过节点，我司不予对单。若后续需要改动提单，则所产生的费用将由贵司承担。"之类的通知函。也就是说，未在规定时间内补料，可能会产生改单费。

◇ **Sample Ⅸ   Okay shipping e-mail**

发件人：XXXXX@XXXXXX.com

Date: PH8KN11 QC report

To: XXXXX

Cc: XXXXX

Dear Limy,

Please find enclosed the QC report of PH8KN11, with my comments below:

- Extra button position: we accept it like this.
- Open seam on side: NOT ACCEPTABLE, please repair or reject pieces.
- Uncut yarn end: NOT ACCEPTABLE, please cut all threads ends.
- Border around the badge must be regular, please check all pieces (picture on page 8).
- Please check all styles: badges must be straight well.
- Packaging looks good thank you.
- Measurements out of our tolerences: we accept because we know on knitting we can't have exact measurements and because it's not a lot of measurements.

You're allowed to ship the goods but only after checking all pieces and rejecting defect one. Thanks in advance for your vigilance.

Best regards

Pascale SCIFO

Adjointe Développement produits/Products development Assistant

■ **Notes：**

1. tolerences: (tolerances)公差度
2. vigilance: 警惕

◇ Sample X Payment for shipper

Dear XXXX,

OBD is 11-JUN-2019.Thanks to provide the bank slip as soon as possible.

此票货已于 6 月 11 号开船，请尽快提供水单以便赎单放货。

Best regards.

XXXX

◇ Sample XI Post the original bill of loading to buyer

Dear XXXX,

昨天我们寄了 6 / 11ETD 的 40 尺柜的正本提单给你，快递单号 SF100178508712,请留意查收，谢谢！

Best regards

XXXXXX

Foreign Trade Documents

◇ **Sample** I **Packing list**

PACKING LIST

1. Shipper/Exporter			8. Invoice No.and date		
YOKI TRADING CO.,LIMITED			YK120701	2012-6-31	
UNIT 04, 7/F, BRIGHT WAY TOWER, NO. 33			9. REMARK		
MONG KOK ROAD, KOWLOON, HK.					
Tel: 00852-23892981	FAX:00852-35902333				
2. For Account & Risk OF Messes			* LC NO:	NO L/C	
SALT AND PEPPER CLOTHING, INC.					
4770 E.50TH ST VERNON, CA 90058			* FABRIC CONTENT :	SELF:100%POLYESTER	
TEL : 323-232-XXX FAX:323-232-XXXX				CONTRAST:50%RAYON	
3. Notify Party					50%NYLON
SALT AND PEPPER CLOTHING, INC.				LINING:100%POLYESTER	
4770 E.50TH ST VERNON, CA 90058					
TEL : 323-232-7XXX FAX:323-232-XXXX					
4. Port of Lolding	5.Final Destintation				
GUANGZHOU,CHINA	LOS ANGELES, CA, USA				
6. Carrier	7. Sailing on or About				
BY SEA	8/25/2012				
10.Marks and Nos. of PKGS	11. Description of Goods		12. Quantity	13. N-WEIGHT 15. G-WEIGHT	16. MEASUREMENT
10 CTN	WOMEN DRESS		986 PCS	170 KG 157 KG	1 CBM
HANDLE WITH CARE					
	PO NO	STYLE NO	ITEAM	QUANTITY CTN	
SALT & PEPPER	SNP86506	JD1759		986PCS 10 CTNS	61X41X38CM
CARTON					
P.O NUMBER					
STYLE NUMBER					
COLOR					
SIZE	TOTAL			986PCS 10 CTNS	
COUNT					
TOTAL PACK					
TOTAL Q'TY					
MADE IN CHINA					
			TOTAL NET WEIGHT :	157 KG	
			TOTAL GROSS WEIGHT :	170 KG	
			TOTAL MEASUREMENT :	1 CBM	
			17. SINEND BY		

◇ Sample Ⅱ Container booking form

SHIPPING ORDER	
Cargo Ready Date can not empty!	

Shipper					**XXXXX HONG KONG LIMITED.**		
Name:	XXX KNITTING CO.,LTD				**Address:**		
Address:	XXXX				**TEL:(852)xxxxxxx FAX:(852)xxxxxxxx**		
					EMAIL: xxxxx@gmail.com		
Tel:	XXXXXX	Fax:	XXXX				
Email:					**Also Notify**		
Contact:					Name:		
Consignee					Address:		
Name:	XXXXXXX						
Address:	XXXXXXX						
					Incoterms:		
					S/O NO.		
Tel:	XXXXXX	Fax:	XXX		**Booking Party**		
Email:	XXXXXXX				Name:		
Contact:	XXXXXX				Address:		
Notify Party							
Name:	XXXXXXXX						
Address:	XXXXXXXXX				Tel:		Fax:
					Excess Declaration	Value	**No. Of Bill Required**
					Service Type		
Tel:	XXXXXXXXX	Fax:					
Email:	XXXXXXX				**Sea Freight To Be**		
Contact:	XXXXXXX						

Vessel Name		Voyage No.	Document Required		
Place Of Receipt		Place Of Delivery			
				Cargo Ready Date	
Port of Loading		Port of Discharge		Final Destination	
XXX					

PARTICULARS DECLARED BY SHIPPER

Cargo Line Item

Item	Marks & NOS	Package(s)	Description of Goods	Gross WT. KGS	Measurement CBM
1	LIF/08/09/2022		LADIES' SWEATER		10.540
2	JAK/39/40/53/2022		LADIES' SWEATER		8.800
3	MAO/44/2022		LADIES' SWEATER		2.000
	Total:				21.340

Container Line Item

Container Size	Carrier SO No.	PKG	PKG Unit	Container No.	Seal No.	KGS	CBM
Booking CY Container QTY		20'	V	40'			
		40'HQ		45'HQ			

Remark	

CFS CLOSING:			仓库 地址	
ETD			交收 时间	
ETA				

All transactions of Herport Hong Kong Limited are subject to its Standard Trading Conditions

(A copy is available upon request), which in certain circumstances limit or exempt its liability.

Neither the carrier nor Herport Hong Kong Limited will assume liability for cargo off loaded or shut out howsoever caused.

**********本托书以邮件方式发送，与原件有同等效力**********

◇ Sample Ⅲ　VGM form

集装箱重量验证信息 (VGM) 申报表

责任方名称和地址		VGM 授权联系人信息	
公司名称：	GUANG DONG XINGYI DA IMP AND EXP CO LTD	姓名：	SASA
地址：	901,B1,DONGGUAN TIAN AN CYBER PARK,NO.1 HUANGJIN,ROAD,NANCHENG, DONGGUAN,GUANGDONG, CHINA.	电话：	18827949910
		电子邮件：	sasa.yao@rongsentextiles.com
城市：	DOGNGUAN	职位：	
邮政编码：	523000		
国家：	CHINA		

BOLLORE LOGISTICS JOBFILE 编号：
订舱编号：ATD0341592

集装箱和 VGM 信息

内部参考 ：	集装箱编号：	集装箱类型：	铅封（如有*）：	VGM：	测量单位（千克或磅）：	VGM 确定方式（1 或 2）**：	确定日期：
	APZU4394072	40GP		11229.8	千克	方法二	2019/6/6

从 2016 年 7 月 1 日起，托运人须按照《国际海上人命安全公约》(SOLAS) 新规提供 VGM 信息。如果托运人未及时提供 VGM，承运商将拒绝集装箱装船，而且如果因为未申报 VGM 或申报不妥而产生任何后果、费用或罚金，相关责任由托运人承担。
本人声明上表所填 VGM 按照 SOLAS 的要求确定。本人授权 Bolloré Logistics 将 VGM 数据传输验证将这承运商。

签名和公司章

* 在公共区域、吸测、称测和中车，集装箱的密封时，是强制付中。
** 方式 1，使用经校准和认可的设备对集装箱整体称重量。
方式 2，使用集装箱完成包装作业所使用同重士智机关夫认可可的方法对集装箱内的所有包装和货品（包括托盘、垫料和系固材料）分别称重量，再加上集装箱的重量。

SOLAS VGM

◇ Sample Ⅳ　SI form

Shipping　Information (SI) 请全部用大写字母输入	
SO NO.	149103098533
FREIGHT COLLECT	
1.Mother VSL/VOG:	CMA CGM CONCORDE 0FL9BW1MA
2.Port of loading:	CNYTN (YANTIAN)
3.Port of discharge:	FRLEH (LE HAVRE)
4.Final destination:	FRLEH (LE HAVRE)
SHIPPER:	
XXXX　CO .,LTD	
ADDRESS: XXXX	
Consignee:	
xxx (Company name)	
ADDRESS:	
Notify:	
SAME AS CONSIGNEE	
MARKS AND NOS	

两个以上 HS CODE,请提供 HS CODE 对应的件重量(请在补料时提供在货物描述上)只有一个 HS CODE 绿色部分的可以不填

DESCRIPTON OF GOODS	H.S. CODE	CTNS	GW(KGS)	REMARKS
LADIES 50% VISCOSE 20%POLYAMIDE 30%POLYESTER PULLOVER (LIF/08/2022)	6110300090	105	273	PULL ACM1101
LADIES 48%VISCOSE 21%POLYAMIDE 31%POLYESTER PULLOVER (LIF/09/2022)	6110300090	209	815.1	PULL ACM0802
LADIES 60%VISCOSE 40%COTTON PULLOVER (DISH/07/2022)	6110300090	8	103.6	ACM1136
LADIES 90%POLYESTER 10%METALLIC FABRIC PULLOVER (GAO/03/2022)	6110300090	21	226	ACM1400/ HSE TGC
LADIES 53%VISCOSE 30%POLYESTER 17%POLYAMIDE PULLOVER (GAO/18/2022)	6110300090	86	224	ACM0507/SU
LADIES 51%VISCOSE 28%POLYESTER 21%POLYAMIDE PULLOVER (JAK/53/2022)	6110300090	30	310.1	440706
LADIES 70%VISCOSE 30% POLYAMIDE PULLOVER (JAK/39/2022)	6110300090	61	296.5	242701
LADIES 70%VISCOSE 30%POLYAMIDE PULLOVER (JAK/40/2022)	6110300090	104	505.4	246701

CONTAINER:	SEAL NO	CTNS	GW (KGS)	CBM	CONTAINER TYPE
EGHU3242886	EMCGLW6421	624	2753.7	24.61	1*20GP
	TOTAL	624	2753.7	24.61	

Issue Original bill or Telex release?	
Issue Feeder bill of Mother vessel bill?	

◇ Sample Ⅴ Container loading

订舱号AJD0341592
柜号APZU4394072
柜重3700KG
货重7529.8KG
柜型40GP
封条K9921892
车牌粤BF●●●
邓●●1353●●289

■ Notes：

封柜资料一般包括订舱号、柜号、柜重、货重、柜型、封条、车牌号和经办人。

◇ Sample Ⅵ　Bill of lading (draft)

NOT NEGOTIABLE UNLESS CONSIGNED TO ORDER　　OCEAN BILL OF LADING

Shipper DONGGUAN HUAYI SUPPLY CHAIN CO ., LTD ROOM 902, UNIT 1,BUILDING2, TIANNAN CYBER PARK , NO .1, HUANGJIN ROAD , NANCHENG STREET , DONGGUAN CITY, GUANGDONG PROVINCE	B/L NO. SZX6023152

Consignee (if 'To Order' so indicate)
SPRINGWAY
3 RUE HENRI POINCARE
93270 SEVRAN
FRANCE

RECEIVED by the Carrier the Goods as specified below in apparent good order and condition unless otherwise stated, to be transported to such place as agreed, authorised or permitted in this Bill of Lading and subject to all the terms and conditions appearing on the front and reverse pages of this Bill of Lading to which the Merchant agrees by accepting this Bill of Lading, any local privileges and customs notwithstanding.

Notify party (No claim shall attach for failure to notify)
SAME AS CONSIGNEE

For delivery of goods please apply to:
HERPORT LE HAVRE
79 RUE JULES SIEGFRIED
76600 LE HAVRE , FRANCE
CONTACT :SOPHIE LEMBERT S.LEMBERT@HERPORT.FR
TEL :02.32.74.61.61 VAT:FR44382329035

Pre-carriage by	Place of receipt YANTIAN
Vessel / Voy No. CMA CGM ZHENG HE/0FM79W1MA	

Port of Loading YANTIAN	Port of discharge LE HAVRE ,FRANCE	Place of delivery LE HAVRE ,FRANCE

Particulars furnished by the merchant

Container No. Marks and Number	Seal No.	Number of Containers or packages	Kind of packages;Description of goods	Gross Weight KGS	Measurement CBM

MARKS AND DESCRIPTION AS PER ATTACHED LIST

_____348 CARTON(S)_____　　　　　　_____2453.10____18.710__

CONTAINER NO.	SEAL NO.	SIZE	MOVEMENT	PKG UNIT	KGS	CBM
TRHU3047166	C0530122	20'GP	CY/CY	348 CARTON(S)	2,453.10	18.710

The particulars of the Goods given above as stated by the shipper and the weight, measure, quantity, condition, contents and value of the Goods are unknown to the Carrier.

FREIGHT COLLECT　　***TELEX RELEASED***

* Total number of containers or other packages or units received by the Carrier(in words)　ONE (1)x20'GP CONTAINER(S) ONLY

Freight and Charges	Prepaid	Collect	Excess limit declaration as per Clause 16
			One(1) original Bill of Lading must be surrendered duly endorsed in exchange for the Goods or delivery order. In WITNESS whereof one(1) original Bill of Lading has been signed if not otherwise stated below, the same being accomplished the other(s), if any,to be void.
Freight payable at DESTINATION	Number of Original B(s)/L THREE (3)		STAMP / SIGNATURE OF THE CARRIER:
LADEN ON BOARD THE VESSEL 　11 Jun 2021 Date	Place and Date of issue SHENZHEN　11 Jun 2021		

JURISDICTION AND LAW CLAUSE
The contract evidenced by or contained in this Bill of Lading is governed by the law of Hong Kong and any claim or dispute under this Bill of Lading or in connection with this Bill of Lading shall be determined exclusively by the Courts in Hong Kong and no other Court.

◇ **Sample Ⅶ Importer security filing**

(For Ocean Export to / via US only)				
IMPORTER SECURITY FILING				
PART I - SHIPMENT DETAIL				
△ 1. SHIPPING ORDER NO.		2. PORT OF LOADING		3. PLACE OF DELIVERY
		XXX		
△ 4. ISF CHARGE (please select either of option)		5. OCEAN VESSEL	**AMS BL# / HOUSE REF#**	
			AMS SCAC	
PART II - SUPPLY CHAIN				
△ 8. SELLER'S NAME & ADDRESS		△ 9. SHIP TO NAME & ADDRESS		
ATTN:	TEL:	ATTN:		TEL:
CITY:	STATE:	CITY:		STATE:
COUNTRY:	POSTAL CODE:	COUNTRY:		POSTAL CODE:
△ 10. BUYER'S NAME & ADDRESS		△ 11. MANUFACTURER (SUPPLIER) NAME & ADDRESS:		
		(pls use the worksheet for additional Manufacturer Information)		
ATTN:	TEL:	ATTN:		TEL:
CITY:	STATE:	CITY:		STATE:
COUNTRY:	POSTAL CODE:	COUNTRY:		POSTAL CODE:
△ 12. CONTAINER STUFFING LOCATION NAME & ADDRESS		△ 13. CONSOLIDATOR NAME & ADDRESS:		
ATTN:	TEL:	ATTN:		TEL:
CITY:	STATE:	CITY:		STATE:
COUNTRY:	POSTAL CODE:	COUNTRY:		POSTAL CODE:
△ 14. IMPORTER NAME & EIN#:		△ 15. CONSIGNEE NAME & EIN#:		
ATTN:	TEL:	ATTN:		TEL:
REG TYPE:		REG TYPE:		
REG No.:		REG No.:		
CITY:	STATE:	CITY:		STATE:
COUNTRY:	POSTAL CODE:	COUNTRY:		POSTAL CODE:
△ 16. BOND TYPE:		△ 17. BOND HOLDER NAME & EIN#:		
△ 18. BOND #:				

PART III - LIST OF ITEM(S)			
△ CONTAINER (size)#	ITEM DESCRIPTION	△ TARIFF CODE (to 6 digits if available)	△Country of Origin

I CERTIFY THAT THIS INFORAMTION IS TRUE AND CORRECT AND AUTHORIZE HERPORT TO PREPARE IMPORTER SECURITY FILING TO US CUSTOMS AND BORDER PROTECTION (CBP) AND OTHER OFFICALS ON THE BASIS OF THIS INFORMATION.

SHIPPER: SIGNATURE: DATE:

△ REQUIRED FIELDS

Oral Practice

Forming a group of two or three partners. Try to work out a dialogue related to the theme of this unit and then perform it in class.

Self-Assessment

Evaluate your practice by marking the following each corresponding item and work out the total scores.

考核情境(分值)	考核要求(分值)	得分
语音语调(20 分)	发音准确，声音清晰(10 分)	
	语调自然，语速自然流畅(10 分)	
内容(30 分)	内容完整(10 分)	
	内容表述符合给定情境(10 分)	
	前后内容逻辑一致，内容之间无矛盾(10 分)	
语言描述(30 分)	用词规范、准确，无语法错误(20 分)	
	表述无歧义(10 分)	
其他(20 分)	无雷同状况(10 分)	
	无严重偏题(10 分)	
合 计(100 分)		

Task 6　Negotiation on Insurance

知识目标

◆ 学习国际贸易中保险谈判的内容
◆ 学习国际贸易中保险谈判的艺术
◆ 学习相关商务沟通的口头英文表达方式
◆ 学习相关商务沟通的书面英文表达技巧

能力目标

◆ 能够准备保险谈判材料
◆ 能够顺利开展保险谈判活动
◆ 能够使用双语顺畅地进行商务口头沟通
◆ 能够处理保险谈判往来信函

Basic Knowledge Review

1. What does negotiation on insurance involve?

Three topics are involved during the negotiation on insurance.

(1) Determine the insurance agencies for the exported goods

International trade insurance indemnifies importers and exporters against various types of losses, including damage to goods in transit, products injuring consumers and importer non-payment. To indemnify means to compensate a company when it loses money due to one of these events. Insurance agencies take on some of the risk so that exporters, particularly, are able to seize opportunities to expand their businesses into foreign markets.

(2) Choose the type of insurance policy

Marine Insurance policy is compulsory in international trade so that all goods passing through the sea must be covered.

Marine Insurance is a type of insurance that covers cargo losses or damage caused to ships, cargo vessels, terminals, and any transport in which goods are transferred or acquired between different points of origin and their final destination. Providing protection against transport-related losses, this voyage policy provides a haven for shipping companies and couriers because it protects them from costly potential losses while transporting goods by water.

(3) Make clear the Institute Cargo Clauses needed to defend against possible losses.

Institute Cargo Clause C provides basic coverage and includes a restricted list of risk covers. It covers the shipment against events such as fire, discharge of cargo in case of distress, explosion, accidents like sinking, capsizing, derailment, collision, etc.

Institute Cargo clause B offers an additional layer of protection. Not only does it include all the risk covers provided under Clause C, but it also covers the shipment against events such as earthquake, volcanic eruption, and damage due to rainwater, seawater, river water, etc., and loss to package overboard or during loading and unloading.

Institute Cargo Clause A provides maximum coverage as it covers all risk of loss or damage to the goods. Apart from the risks covered under Clauses B and C, it also covers losses due to breakage, chipping, denting, bruising, theft, non-delivery, all water damage, etc.

Risks such as wars, strikes, riots, and civil commotions are not covered under the institute cargo clauses. However, the insurer may provide this cover on payment of additional marine insurance premium.

So in terms & conditions of marine insurance coverage, these three types of marine insurance clauses: Institute Cargo Clauses A, B, and C. Clause A provides maximum coverage, Clause C provides basic risk coverage.

2. What is not covered under Marine Insurance?
✓   Delivery issues
✓   Renovation and repairs
✓   Bad quality goods
✓   Intentional loss
✓   Personal insolvency
✓   Wars and situations

3. What is a cargo insurance policy?
Insurance policy is a legally binding written document, which is issued by insurance company or underwriter to policy holder or insured/assured.

Just like bills of lading (bill of lading states terms and condition of the carriage as a transport document), insurance policies define the terms and conditions of the insurance contract and serves as a legal evidence of the insurance agreement.

International marine cargo insurance policies are generally issued subject to ICC Cargo Clauses such as Institute Cargo Clauses A, Institute Cargo Clauses B and Institute Cargo Clauses C.

An insurance cargo policy should specify the terms on which the indemnity cover has been provided by giving an express reference to one of above mentioned ICC cargo clauses.

Furthermore, an insurance cargo policy may mention additional risks covered such as war risks and strike risks.

In some occasions, insurance cargo policies indicate that the cover is subject to a franchise or excess (deductible).

Other information that could be mentioned on cargo insurance policies are:
- ✓ Amount of insurance premium
- ✓ Shipment details such as port of loading, port of discharge, vessel name and voyage number, description of goods etc.
- ✓ Currency of the insurance cover
- ✓ Insurance cover amount
- ✓ Insurance company's agent at the port of destination
- ✓ Procedures for the claim and required documents
- ✓ Important aspects of insurance policies
- ✓ Insurance policies issued by the insurance companies for specific transactions. An insurance company gets the details of the shipment from the policy holder and prepares his insurance policy offer.

If insurance holder agrees on the terms and conditions as well as insurance premium, then the insurance company or the underwriter issues the insurance policy.

An insurance policy generally issued when the goods are loading and expires on completion of unloading from the carrying vessel at the port of destination.

📝 Task Descriptions

Negotiation on Insurance

　　杭州红粉服饰有限公司代表与美国斯特朗服装有限公司代表就价格、运输谈判成功之后，开始进入产品保险相关问题的谈判。

　　接下来的对话就是杭州红粉服饰有限公司代表与美国斯特朗服装有限公司代表围绕产品保险谈判等一系列活动展开的。

🎧 Typical Dialogues

◇ Dialogue Ⅰ

　　这段对话发生在杭州红粉服饰有限公司外贸业务员 Dicky 与中国人民保险公司国际业务部客户经理李明之间，主要讨论国际海洋运输途中货物保险的类型。

A：杭州红粉服饰有限公司外贸业务员 Dicky

B：中国人民保险公司国际业务部客户经理李明

A：Where the kinds of insurance are concerned, I am a layman. Could you please explain them to me?

A：对这些险种，我是个外行。您能跟我解释一下吗？

B：Of course.

B：当然可以。

A：There are risks like F.P.A. and W.P.A. I know that F.P.A. stands for "Free from Particular Average", while W.P.A. stands for "With Particular Average". That means W.P.A. covers more risks than F.P.A, doesn't it?

A：平安险和水渍险。我知道，F.P.A.代表"平安险"，W.P.A.代表"水渍险"。水渍险比平安险承保的范围更大，对吗？

B：Yes, The W.P.A. clause covers certain percentage of the nature of particular average, while F.P.A. doesn't. F.P.A. can be considered as a limited clause.

B：是的。水渍险条款比平安险多承保了一定百分比的单独海损性质的损失，平安险承保的范围还是很有限的。

A：Do you mean that it does not cover risks like theft and pilferage, freshwater, oil and contamination, etc.?

A：您的意思是说平安险不包括盗窃和偷窃、淡水、油、污染等风险？

B：You are right. Those risks have to be specially applied for.

B：对的。这些风险需要另外申请。

A：Then which side of the business will pay for these special risks?

A：那么特别险应由哪方企业支付呢？

B：Well, usually it is the buyer's account for the additional premium. The seller insures against such risks if the buyer requires, but the price will go up. Unless buyers note clearly in the L/C what kinds of insurance they'd like to cover, prices are generally calculated without insurance for any special risks.

B：通常买家支付额外保费。如果买方要求，卖方会投保特别险，但产品价格上涨。除非买家在信用证中明确表明投保的险种，通常卖家报价是不包含特别险的费用的。

A：Well, I must say your explanations are really very helpful, Now I'm clear about these terms and concepts. Thank you very much.

A：我必须说您的解释对我的确大有帮助。现在我对这些术语和概念清楚多了。非常感谢。

◇ Dialogue Ⅱ

这段对话发生在杭州红粉服饰有限公司外贸业务员 Dicky 与美国斯特朗服装有限公司采购部经理 Betty 之间，主要讨论货物投保的相关问题。

A：杭州红粉服饰有限公司外贸业务员 Dicky

B：美国斯特朗服装有限公司采购部经理 Betty

A：Good morning, Miss Betty, I think we should discuss the terms of insurance for our new contract.

A：早上好，贝蒂小姐，我想我们应该讨论一下新合同的保险条款。

B：Yes, of course. I think we should cover our goods again All Risks so that we can avoid all the risks.

B：当然。我认为我们应该投一切险，这样我们就可以避免所有的风险。

A：All Risks does not mean all the risks. All Risks include W.P.A. (with particular average), F.P.A. (free of particular average), T.P.N.D. (Theft, Pilferage & Non-delivery), fresh water rain damage, risk of shortage, risk of intermixture and contamination, leakage risk, clashing and breakage risk, hook damage, loss and damage by breakage of packing, rusting risk, etc. War risk is not included.

A：一切险并不是所有的风险。一切险包括水渍险、平安险、偷窃及提货不着险、淡水雨淋险、短量险、混杂玷污险、渗漏险、碰损破碎险、钩损险、包装破裂险、锈损险等。战争险不包括在内。

B：I think we have to choose the right insurance coverage to protect our goods.

B：我认为我们必须选择合适的保险来保护我们的货物。

A：We generally insure W.P.A. (with particular average) on a CIF (cost, insurance and freight) offer. Special risks can only be covered upon request.

A：CIF 报价一般投水渍险。特别险要另外申请。

B：What about SRCC(Strike，Riot，and Civil Commotion)? Can we request you to cover this for our imports?

B：罢工、暴动、民变险怎么样？我们要求投 SRCC 可以吗？

A：Yes, we accept it.

A：好的。

◇ Dialogue Ⅲ

　　这段对话发生在杭州红粉服饰有限公司外贸业务员 Dicky 与美国斯特朗服装有限公司采购部经理 Betty 之间，由于 CIF 价成交，由卖方办理保险手续，所以双方就最后的保险事宜进行最后的确认。

A：杭州红粉服饰有限公司外贸业务员 Dicky

B：美国斯特朗服装有限公司采购部经理 Betty

A：How are you going to arrange insurance?

A：你们打算怎么投保？

B：We will insure the goods against W.P.A. at invoice value plus 10%.

B：按发票金额的 110% 投保水渍险。

A：I suggest risk of breakage should be added, for some goods are fragile.

A：有些货易碎，建议再加个破碎险吧。

B：That needs extra premium which should be charged to our account.

B：附加险需另外收费，费用由我方承担。

A：Ok. We will send you the insurance policy in due time.

A：嗯，保险手续办完会及时寄给你们的。

B：Thank you!

B：谢谢。

◇ **Dialogue Ⅳ**

这段对话发生在杭州红粉服饰有限公司外贸业务员 Dicky 与美国斯特朗服装有限公司采购部经理 Betty 之间，主要讨论货物在运输途中发生意外后的相关问题。

A：美国斯特朗服装有限公司采购部经理 Betty

B：杭州红粉服饰有限公司外贸业务员 Dicky

A：Do you remember the order two months ago.

A：还记得两个月前的订单吗？

B：Is everything going well?

B：一切顺利吗？

A：The consignment arrived at the destination yesterday was seriously damaged. The loss through breakage was over 50% of the consignment.

A：昨天货到了，但 50% 损坏了。

B：I am sorry to hear that. Have you applied for a claim to the insurance company?

B：听到这个消息，我很难过。向保险公司提出理赔了吗？

A：Yes, we have. But the insurance company refused to admit liability.

A：是的。但保险公司不受理。

B：What is the point?

B：为什么？

A：They said there was no insurance on breakage. We naturally were not satisfied with such a reply from insurance company. I should like to hear what you have to say about this insurance issue.

A：他们说没有破损险。对保险公司这样的答复我们当然不满意。我想听听你方对这个保险问题的看法。

B：You know of course that as the sellers, we are mediators.

B：你知道，作为卖家我们只是中间人。

◇ **Dialogue Ⅴ**

这段对话发生在美国斯特朗服装有限公司采购部经理 Betty 与中国人民保险公司国际业务部客户经理李明之间，主要讨论货物理赔事宜。

A：美国斯特朗服装有限公司采购部经理 Betty

B：中国人民保险公司国际业务部客户经理李明

A：Hello, Mr. Li. Have you received the surveyor's report?

A：您好，李先生。调查报告收到了吗？

B：Yes, Miss Betty. Through a thorough examination and a careful inspection, twenty cartons of goods have been found seriously damaged. Inside these damaged cartons, the majority of the dresses contained have been water-stained or severely soiled...Oh, what a great loss!

B：收到了，贝蒂小姐。经过彻底地检查，发现有二十箱货物严重损坏。这些损坏纸箱里的衣服都被海水浸透了。损失太大了！

A：Indeed, it is. According to the description of the letter I received, the portion of dresses can not be sold.

A：是的。根据信上描述的情况，那部分衣服根本卖不出去。

B：I trust so. It appears that the damage was caused sometime during the transshipment.

B：我相信。货损应该在转运途中造成的。

A：That might be true. But the packing in some of the cartons was not in accordance with the contract stipulations. According to the contract, each of the cartons should be reinforced with double metal straps.

A：这可能是真的。有些纸箱的包装不符合合同规定。根据合同规定，每个纸箱都应该用双层金属带进行加固。

B：If so, we agree to settle your claims. But that should be done after negotiation with the shipping company, OK?

B：如果是这样，我们同意理赔。具体事宜跟船运公司谈判之后再决定好吗？

A：OK, But what are we to do with that part of goods damaged? Should we return them to you, or hold them at your disposal?

A：好的，但是我们该怎么处理那部分损坏的货物呢？我们是应该把它们给保险公司，还是放我们这，等着你们来处理？

B：I'll give you a definite answer after I consult with the manufacturer.

B：我跟厂商核实情况后，再给你们明确答复吧。

👫 Roles Simulation

Suppose you are one of them in the following conversation. Try to read aloud and practice the underlined sentences.

这段对话发生在美国斯特朗服装有限公司采购部经理 Betty、杭州红粉服饰有限公司外贸业务员 Dicky 与中国人民保险公司国际业务部客户经理李明之间。三方针对海上损失货物的理赔事宜进行商讨。

A：美国斯特朗服装有限公司采购部经理 Betty

B：中国人民保险公司国际业务部客户经理李明

C：杭州红粉服饰有限公司外贸业务员 Dicky

(Asking for the procedure of compensation)

(询问理赔程序)

A: Good morning，Mr. Li. I wonder what is the procedure for filing a claim in the event of loss or damage to my goods.

A：早上好，李经理。我想知道如果货物发生丢失或损坏，理赔程序是怎么样？

B: Hello, Miss Betty. The most commonly adopted practice is that you lodge a claim with the agent of the insurance company at your port within 60 days after arrival of the goods.

B：您好，贝蒂女士。一般来讲，货物到达目的地后 60 天之内，向你方所在地的保险公司代理提出索赔申请。

A：And then the insurance company will compensate me, I think?

A：然后保险公司会赔偿我的损失，对吗？

B：You will first of all have a survey report to support your claim. And <u>the survey report is to be put in within 60 days after the arrival of the consignment. If the loss or damage belongs in the responsibility of the insurance company, they will, in the light of the actual finding, undertake to compensate you according to the risk insured.</u>

B：首先需要投交理赔的事故调查报告。<u>调查报告在货物到达目的地后 60 天内提交。属于保险公司承担责任的，将根据实际损失和相应投保险别进行赔偿。</u>

(Discussing the damaged facts and figures)

(讨论受损情况)

A：Hi, Dicky. <u>the December consignment arrived at our port was seriously damaged.</u> More than 30 percent of the glass dress accessaries were broken.

A：Dicky, <u>12 月份到达我方港口的货物严重受损了。</u>超过 30%的玻璃配饰碎了。

C：It's impossible. Our goods are well-examined before shipping.

C：不可能。货物在装运前我们都严格检验过的。

A：Maybe the losses occurred in transit. <u>Since we concluded this business on CIF basis, I have to claim with you for the losses.</u>

A：也许破损是在运输过程中发生的。<u>我们这单生意是以 CIF 到岸价格条件成交的，这次损失我们只好向贵方提出索赔要求了。</u>

C：Sorry. We cannot pay your claim. It is not our fault. We are not liable for the damage. <u>Owing to the Risk of Breakage is not mentioned in the contract, the losses are not covered.</u>

C：很抱歉，我们不能接受您的索赔要求。这不是我们的过错，我们没有义务对此负责任。<u>合同上并未提及破碎险，这个损失不在你们投保的险种之内。</u>

A：Then who should be responsible for the loss, the insurance company or the shipping company?

A：那该由谁来承担责任呢，保险公司还是运输公司？

C：We suggest that you deal with the insurance company and the shipping company. <u>If you had insured against breakage, it is clearly a case for the insurance company. You can file the claim with the insurance company.</u>

C：我们建议你们与运输公司和保险公司联系。<u>如果你们投了破碎险，这显然是保险公司的责任。你可以向保险公司索赔。</u>

(Solving the losses among three parties)

(三方解决问题)

C：Shall we get down to business right now?

C：我们可以开始谈正事了吗？

A：We naturally were not satisfied with the reply from your company, Mr. Li. We presume the wording of our L/C implies covering the risk of breakage. The loss through breakage was over 30% of the consignment. And you refused to admit liability. Can you explain it to my satisfaction, Mr. Li?

A：李经理，我方对你的答复当然不满意。我认为信用证里包括了破碎险。超过了 30%的货物受损，你们公司拒绝承担责任。你能给我满意的答复吗？

B：In my opinion, <u>the insurance company is responsible for the claim, as far as it is within the scope of coverage.</u> The point is that the loss in question was beyond the coverage granted by us.

B：<u>只要在保险责任范围内，保险公司就应负赔偿责任。</u>问题是你们说的损失并不在我们承保的责任范围。

C：Usually, <u>we applied only for W.P.A. coverage and let our customers deal with the matter of special risk.</u>

C：<u>一般我们给货物投水渍险。特殊险需要客户自己投保。</u>

A：But when I take a W.P.A. insurance, that is, With Particular Average, I should think the risk of breakage is a particular average, isn't it?

A：我们投了水渍险。我以为里面包括破碎险。

B：Not every breakage is a particular average. <u>It is a particular average when the breakage results from natural calamities or maritime accidents, such as stranding and sinking of the carrying vessel, fire, explosion or collision.</u> If none of these conditions occurs, the loss is outside the scope of coverage.

B：不是所有的破碎都属于单独海损。<u>只有自然灾害和意外事故所造成的破损才属于单独海损。比如货船的搁浅和沉没、火灾、爆炸或碰撞。</u>由于所承保货物内在的缺陷或特征所引起的损失，不在承保范围之内。

A：But in the letter of credit, we request coverage for "all marine risks". The risk of breakage is covered by marine insurance, isn't it?

A：信用证中我们要求承保"一切险"。破碎险也应包括在内，不是吗？

B：Of course. <u>But it is the usual practice to make specific mention in the insurance policy or certificate that the risk of breakage is included.</u>

B：当然。<u>可是按照通常的惯例要在保险单或保险凭证上特别注明"破碎险"在内。</u>

A：I must say that you have corrected my ideas about the insurance. I see now that it is far more complicated than I ever imagined.

A：我必须说，你纠正了我对保险的看法。我现在才明白，保险要比我想象得复杂得多。

C：What can we do now?

C：我们能做些什么呢？

A：To compensate a part of the loss, may I ask you to make us a firm offer for 5,000 pieces of similar products CIF London including the risk of breakage, March shipment?

A：为了补偿部分损失，我方能否请你方再提供 5000 件同质物品，CIF 伦敦投破碎险 3 月出货？

C：OK, We will make you an offer tomorrow.

C：好吧，明天给你报实价。

❤❤ *Typical Expressions*

1. We can cover various kinds of insurance, including F.P.A.(Free from Particular Average),

W.P.A.(With Particular Average), A.R.(All Risks), T.L.O.(Total Loss Only), etc. 我们可保各种险别，包括平安险、水渍险、综合险、全损险等。

2. The risks of theft and pilferage, freshwater, oil grease, hooks, leakage, contamination, deterioration, etc. are specifically mentioned in the insurance certificates and must be specifically applied for. 偷盗险、淡水雨淋险、油渍险、钩损险、破碎险、渗漏险、污染险、变质险等在保险单中被特别提出来，所以必须单独申请。

3. We'd like to cover W.P.A. for 110% of the invoice value. 我们想按发票总金额的 110%投保水渍险。

4. The insurance is needed from September 15. 保险需从 9 月 15 号开始投保。

5. Please hold us covered on the cargo listed on the attached sheet. 请为清单上的货物办理投保手续。

6. The rate of the Risk of Breakage is 3%. 破碎险费率是 3%。

7. The loss can be covered within the period of insurance. 承保期内的损失可以得到补偿。

8. Unless we hear from you to the contrary, we shall arrange the insurance on the cargo against the usual risks, for the value of the goods plus freight. 除非收到你们另外的通知，否则我们将根据你们的货物和运费为这批货安排常规险种。

9. In the letter of credit, only coverage for "all marine risks" was requested. 信用证中只要求保"一切海洋运输货物险"。

10. The insurance company refused to admit liability as there was no insurance on breakage. 保险公司拒绝承担理赔责任，理由是没有投保破碎险。

Foreign Trade Letter

◇ **Sample Insurance**

Dear Betty，
Referring to your letter of March 1, 2021 in which you required about CIF insurance for dresses that we mentioned to you on February 28，we provide you with the information as follows.

As to the CIF Deal，we usually cover the insurance against all risks to the People's Insurance Company of China per the relevant Ocean Marine Cargo Clause of the People's Insurance Company of China dated Jan 1,1981. If you claim ICC clauses for insurance, we would be very glad to satisfy your demand on condition that the margin of expenses will be paid by you.

If you like, we could also provide additional coverage for goods with the extra cost on your account. In such case, we will send you a receipt of cost issued by the relevant insurer. As usual the amount insured is 110% of invoice value. If you ask for higher percentage, we could arrange accordingly provided that the extra cost is at your expense.

We hope the above-mentioned information is of interest to you and we are looking forward to receiving your reply as soon as possible.

Your faithfully,

Dicky

Foreign Trade Document

◇ **Sample   Insurance policy**

中国人民财产保险股份有限公司
(PICC Property and Casualty Company Limited)
出口货物运输保险投保单
(APPLICATION FORM FOR EXPORTING CARGO TRANSPORTATION INSURANCE)

投保日期(Date) 20210315

发票号码(Invoice NO.)： 70JJJJKHJLLL 合同号 Contract №.： JJKLLPPPOJ11113 信用证号 L/C №.： JIQQQPPIU122			投保条款和险别 (Insurance clauses and risks)
A statement Insurance is required on the following commodities:　兹有下列物品投保：			PICC CLAUS 中国人民保险公司保险条款
唛头 MARKS & NOS.	包装及其数量 packing and quantity	货物描述 description of goods	(　) ICC CLAUSE 英国协会货物险条款
London 伦敦 CT5132 Totally 500 cartons 总共 500 箱	500 cartons Each set packed in one carton 500 cartons transported in one 40ft container. 500 箱 每个集装一箱， 500 箱运输一个 40 英尺集装箱	女式衬衫 Women's dresses	(√) ALL RISKS 一切险 (　) W.A.水渍险 (　) F.P.A.平安险 (　) WAR RISKS 战争险 (　) S.R.C.C.罢工、暴动、民变险 (　) ICC Clause B 英国协会货物险条款 B
保险金额(Insured amount)	USD ($5,000,000,000.00)		(　) ICC Clause C 英国协会货物险条款 C
起运港(The Loading Port)	(Beilun Port) Ningbo		(　) Air TPT All Risk 航空运输综合险
开航日期 (Date of Commencement)	2021.03.20	船名 (Conveyance)	SUNNY	(　) Air TPT Risk 航空运输险
转内陆 Via	UPTO (Hong Kong, China)		(　) O/L TPT All Risk 陆路运输综合险
目的港(Destination)	(London, UK)		(　) O/L TPT Risk 陆路运输险
赔款地点 (Claims Payable At)	(Ningbo, China)		(　) Transshipment Risks 转运险
赔付币种 (Claims Payable At)	(USD) in the currency of draft (credit)		(　) W/W 仓至仓条款

保单份数 (Original No.s)	(2＋1)张	（　）TPND 偷窃提货不着险 （　）FREC 火险责任扩展条款 （　）IOP 无免赔率 （√）RFWD 淡水雨淋险 （√）Risk of Breakage 破碎险
其他特别条款 (Other Clauses)	THE INSURED CONFIRMS HEREWITH THE TERMS AND CONDITIONS OF THESE INSURANCE CONTRACT FULLY UNDERSTOOD 被保险人确认本保险合同条款和内容已经完全了解。 The interpretation of this proposal shall be subject to English version. 本投保单内容以英文为准。 Only the written form contract will be operated, any other form will be not approved. 本保险合同一律采用书面形式，双方不认可其他形式的约定。 This insurance contract will be effective when the policy is issued by the underwriter and when the insurance premium is received according to the terms of the contract by this company. 本保险合同自保险人核保并签发保险单后成立，自投保人依约缴费后生效，保险人自本保险合同生效后开始承担保险责任。 In the event of any dispute arising from its implementation or enforcement, either of the parties to the Contract of Insurance may make application to the China International Arbitration Committee, whose judgements shall be given in accordance with such rules of arbitration as are then in effect. 因履行保险合同发生争议的，一方可向中国国际仲裁委员会依该会届时有效的仲裁规则申请仲裁。	

Together With The Following Documents   随附产品资料
(1) Manual 产品说明书 (2) Certification of Quality 质量合格证书 (3) Safety Warning Mark 安全警告标记 (4) License (s) 许可证 (5) Quality Inspection Report 质量检验报告 (6) Sales Contract 销售合同 (7) Design Drawing 设计图纸 (8) Else 其他

以下由保险公司填写 Following insurance companies to fill			
保单号码(Proposal №.)	0104324562159	签单日期(Date)	2021.03.18

投保人(The Insured)

杭州红粉服饰有限公司

Hangzhou Pinklife Fashion co. Ltd.

A⊘C Oral Practice

Forming a group of two or three partners. Try to work out a dialogue related to the theme of this unit and then perform it in class.

Self-Assessment

Evaluate your practice by marking the following each corresponding item and work out the total scores.

考核情境(分值)	考核要求(分值)	得分
语音语调(20 分)	发音准确，声音清晰(10 分)	
	语调自然，语速自然流畅(10 分)	
内容(30 分)	内容完整(10 分)	
	内容表述符合给定情境(10 分)	
	前后内容逻辑一致，内容之间无矛盾(10 分)	
语言描述(30 分)	用词规范、准确，无语法错误(20 分)	
	表述无歧义(10 分)	
其他(20 分)	无雷同状况(10 分)	
	无严重偏题(10 分)	
合　　计(100 分)		

Task 7   Negotiation on Payment

知识目标

◆ 学习国际贸易中支付谈判的内容
◆ 学习国际贸易中支付谈判的艺术
◆ 学习相关商务沟通的口头英文表达方式
◆ 学习相关商务沟通的书面英文表达技巧

能力目标

◆ 能够准备支付谈判材料
◆ 能够顺利开展支付谈判活动
◆ 能够使用双语顺畅地进行商务口头沟通
◆ 能够处理支付谈判往来信函

Basic Knowledge Review

1. What are methods of payment in international trade?

International payments, also known as cross border payments or global payments, are transactions that involve more than just banks. They connect companies, individuals, banks, and settlement institutions operating in at least two different countries with different currencies that need to be paid.

As shown in *the following figure*, there are five primary methods of payment for international transactions. During or before contract negotiations, you should consider which method in the figure is mutually desirable for you and your customer.

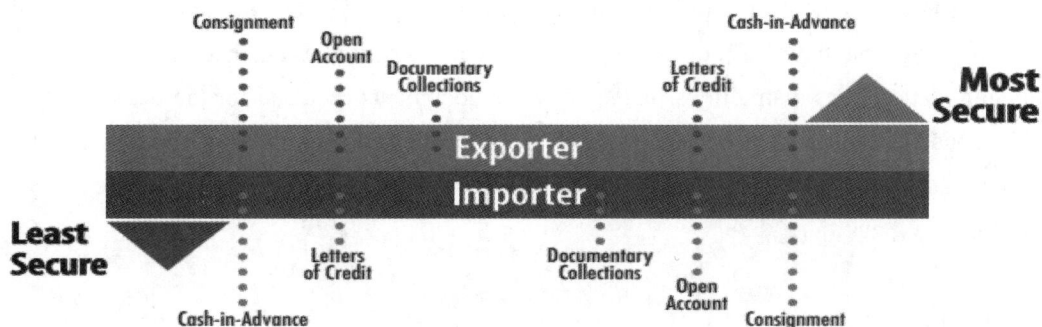

(1) Cash-in-Advance

With cash-in-advance payment terms, an exporter can avoid credit risk because payment is received before the ownership of the goods is transferred. For international sales, wire transfers and credit cards are the most commonly used cash-in-advance options available to exporters. With the advancement of the Internet, escrow services are becoming another cash-in-advance option for small export transactions. However, requiring payment in advance is the least attractive option for the buyer, because it creates unfavorable cash flow. Foreign buyers are also concerned that the goods may not be sent if payment is made in advance. Thus, exporters who insist on this payment method as their sole manner of doing business may lose to competitors who offer more attractive payment terms.

(2) Letters of Credit

Letters of credit (LCs) are one of the most secure instruments available to international traders. An LC is a commitment by a bank on behalf of the buyer that payment will be made to the exporter, provided that the terms and conditions stated in the LC have been met, as verified through the presentation of all required documents. The buyer establishes credit and pays his or her bank to render this service. An LC is useful when reliable credit information about a foreign buyer is difficult to obtain, but the exporter is satisfied with the creditworthiness of the buyer's foreign bank. An LC also protects the buyer since no payment obligation arises until the goods have been shipped as promised.

(3) Documentary collections

A documentary collection (D/C) is a transaction whereby the exporter entrusts the collection of the payment for a sale to its bank (remitting bank), which sends the documents that its buyer needs to the importer's bank (collecting bank), with instructions to release the documents to the buyer for payment. Funds are received from the importer and remitted to the exporter through the banks involved in the collection in exchange for those documents. D/Cs involve using a draft that requires the importer to pay the face amount either at sight (document against payment) or on a specified date (document against acceptance). The collection letter gives instructions that specify the documents required for the transfer of title to the goods. Although banks do act as facilitators for their clients, D/Cs offer no verification process and limited recourse in the event of non-payment. D/Cs are generally less expensive than LCs.

(4) Open account

An open account transaction is a sale where the goods are shipped and delivered before payment is due, which in international sales is typically in 30, 60 or 90 days. Obviously, this is one of the most advantageous options to the importer in terms of cash flow and cost, but it is consequently one of the highest risk options for an exporter. Because of intense competition in export markets, foreign buyers often press exporters for open account terms since the extension of credit by the seller to the buyer is more common abroad. Therefore, exporters who are reluctant to extend credit may lose a sale to their competitors. Exporters can offer competitive open account terms while substantially mitigating the risk of non-payment by using one or more of the appropriate trade finance techniques covered later in this guide. When offering open account terms, the exporter can seek extra protection using export credit insurance.

(5) Consignment

Consignment in international trade is a variation of open account in which payment is sent to the exporter only after the goods have been sold by the foreign distributor to the end customer. An international consignment transaction is based on a contractual arrangement in which the foreign distributor receives, manages, and sells the goods for the exporter who retains title to the goods until they are sold. Clearly, exporting on consignment is very risky as the exporter is not guaranteed any payment and its goods are in a foreign country in the hands of an independent distributor or agent. Consignment helps exporters become more competitive on the basis of better availability and faster delivery of goods. Selling on consignment can also help exporters reduce the direct costs of storing and managing inventory. The key to success in exporting on consignment is to partner with a reputable and trustworthy foreign distributor or a third-party logistics provider. Appropriate insurance should be in place to cover consigned goods in transit or in possession of a foreign distributor as well as to mitigate the risk of non-payment.

2. What is Letter of Credit (L/C)?

A letter of credit (L/C), also known as a documentary credit or bankers commercial credit, or letter of undertaking (L/U), is a payment mechanism used in international trade to provide an economic guarantee from a creditworthy bank to an exporter of goods. Letters of credit are used extensively in the financing of international trade, where the reliability of contracting parties cannot be readily and easily determined. Its economic effect is to introduce a bank as an underwriter, where it assumes the counterparty risk of the buyer paying the seller for goods.

3. How do traders run a Letter of Credit (L/C)?

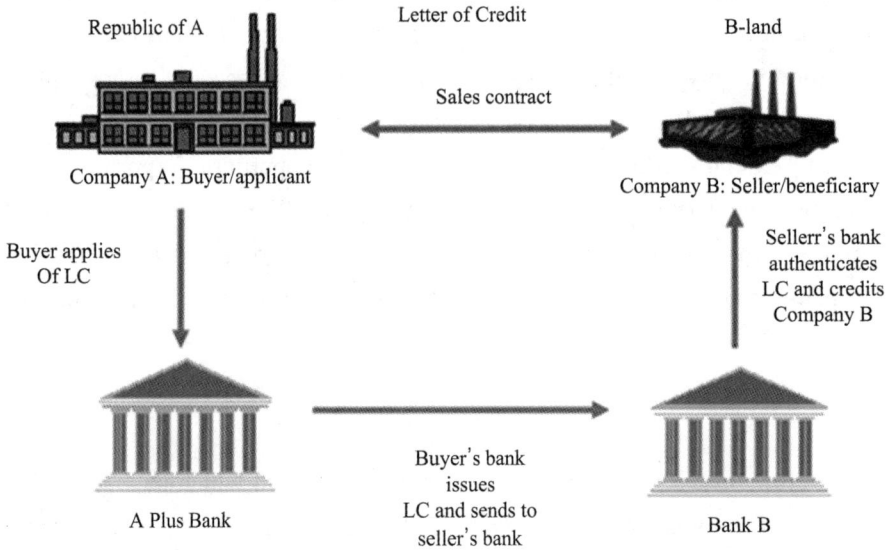

Republic of A

Letter of Credit

B-land

Sales contract

Company A: Buyer/applicant

Company B: Seller/beneficiary

Buyer applies
Of LC

Sellerr's bank
authenticates
LC and credits
Company B

A Plus Bank

Buyer's bank
issues
LC and sends to
seller's bank

Bank B

4. What is the procedure of L/C Documentary Collection?

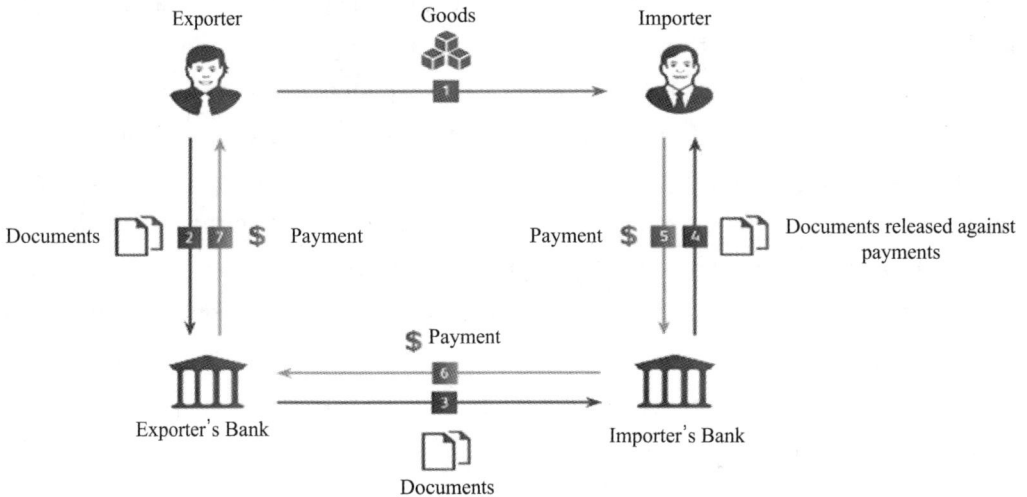

Exporter

Goods

Importer

1

Documents

2 7 $ Payment

Payment $ 5 4

Documents released against payments

$ Payment

6

3

Exporter's Bank

Importer's Bank

Documents

5. What documents should be offered to the freight forwarder while proceeding L/C ?

Sellers usually should offer the following documents.

(1) Commercial invoice: three originals, signed and sealed;

(2) Packing list: three originals, signed and sealed, informing the information of the number of boxes, weight, and clothes size of each box;

(3) Bill of Lading: three originals, signed and sealed by the freight forwarder, remarking freight collect, L/C account number and date of departure;

(4) Certificate of origin: one original and one copy

(5) SGS report signed and approved by the customer: a copy of the first page of the SGS report, provided by the customer.

6. Letter of credit sample format

Usually, we use irrevocable letter or credit in international trade.

This format is for use in designing a Letter of Credit Instructions form appropriate for your own company.

LETTER OF CREDIT INSTRUCTIONS

Date: _____

To:

From	_____
Address	_____
City & State	_____
Country	_____ Zip Code _____
Attn	_____
Telephone	_____
Fax	_____

RE: ☐ Our Pro-Forma Invoice# _____ Dated _____
       ☐ Your Purchase Order# _____ Dated _____
       ☐ Commercial Contract# _____ Dated _____

Gentlemen:

In connection with your above-referenced purchase, the following terms and conditions are for inclusion in your irrevocable letter of credit. We are providing you with these details as a confirmation of our understanding of the terms of sale covering this transaction. If these details do not agree with your understanding or if you are unable to comply with these terms and conditions, please notify us prior to the issuance of your letter of credit to avoid unnecessary delays and costs. Thank you for your patronage and cooperation.

1. The letter of credit must be issued no later than _____ by a bank acceptable to us.

2. The letter of credit must be irrevocable and be subject to the 2007 Revision of the Uniform Customs and Practice for Documentary Credits published by the International Chamber of Commerce (UCP600).

3. The letter of credit must state that it is available with any bank by negotiation.

4. The letter of credit must be opened with full details by SWIFT

       In favor of: _____   [indicate the company name and address you will use
                    _____   in your invoices; if this is not the address you want
       Attn: _____   your L/Cs mailed to, give separate instructions for
       Telephone: _____   where this L/C is to be sent]

       We will not initiate shipment until the actual letter of credit is received but it may expedite processing if you will fax a copy of the letter of credit to [name] at [e-mail address]. This must be a copy of your bank's actual SWIFT message sent to the advising bank. A copy of your letter of credit application is not sufficient.

5. The letter of credit must be payable in U.S. dollars for
       ☐ up to an amount of _____
       ☐ an approximate amount of _____

✎ **Task Descriptions**

Negotiation on
Payment

　　杭州红粉服饰有限公司代表与美国斯特朗服装有限公司代表就价格、运输、保险谈判成功之后，开始进入产品支付相关问题的谈判。

　　接下来的对话就是杭州红粉服饰有限公司代表与美国斯特朗服装有限公司代表围绕产品支付谈判等一系列活动展开的。

🎵 **Typical Dialogues**

◇ **Dialogue** Ⅰ

　　这段对话发生在杭州红粉服饰有限公司外贸业务员 Dicky 与美国斯特朗服装有限公司采购部经理 Betty 之间，双方讨论支付细节。

A：美国斯特朗服装有限公司采购部经理 Betty

B：杭州红粉服饰有限公司外贸业务员 Dicky

A：Now, Let's talk about the terms of payment. Do you accept D/A or D/P?

A：现在谈谈支付方式吧。贵公司接受承兑交单或者付款交单吗？

B：I'm sorry to say that we only accept L/C.

B：很抱歉，我们只接受信用证。

A：As you know, the western market has been declining recently. Business is not easy as it used to be. As an old client of yours, I think we should enjoy your special treatment. I hope you would allow us to pay by D/A or D/P.

A：您知道的西方市场一直处于下滑状态。生意已不似从前了。我是你们的老客户，这点优待可以吗？我希望你们能允许承兑交单或者付款交单。

B：I understand your situation. As you pointed out, the western economy is going down, and the international financial market is not stable. To be on the safe side, we can't make exceptions.

B：您的处境我们能理解。正如您所指出的，西方经济下滑，国际金融市场不稳定。为了安全起见，我们不能破例。

A：It will increase our expenses to open the L/C and tie up our funds.

A：开信用证会增加费用，占用我方资金。

B：Since you are our old customer and your order is quite large, how about 70% by L/C, 30% by D/P? This is not our normal practice.

B：鉴于您是我们的老客户，而且订单也大，信用证付 70%，付款交单付 30%可以吗？这不是我们惯用的做法。

A：All right, I agree.

A：好吧，我同意。

◇ Dialogue Ⅱ

这段对话发生在杭州红粉服饰有限公司外贸业务员 Dicky 与美国斯特朗服装有限公司采购部经理 Betty 之间，卖家催买家开立信用证。

A：美国斯特朗服装有限公司采购部经理 Betty

B：杭州红粉服饰有限公司外贸业务员 Dicky

A：Could you make sure that the goods will be delivered before May so that they can catch up with the sales season before the Children's Day?

A：为赶上儿童节的销售旺季，你方能确保在 5 月之前交货吗？

B：In this case, you'd better open the confirmed irrevocable L/C payable against documents before April 5. Since we need time to get the goods ready and book the shipping space, I suggest to stipulate the time of shipment as within 14 days after the receipt of the L/C.

B：那最好在 4 月 5 日前开好承兑交单不可撤销的信用证，因为我们需要时间备货订舱。我建议装运时间为收到信用证后 14 天内。

A：That's fine.

A：好的

◇ Dialogue Ⅲ

这段对话发生在杭州红粉服饰有限公司外贸业务员 Dicky 与美国斯特朗服装有限公司采购部经理 Betty 之间，双方讨论使用人民币支付的问题。

A：杭州红粉服饰有限公司外贸业务员 Dicky

B：美国斯特朗服装有限公司采购部经理 Betty

A：Shall we talk about payment in Chinese currency? You know, many of our business friends are paying for our exports in Renminbi.

A：谈谈用人民币付款好吗？许多欧洲朋友用人民币与我们做生意。

B：I know some of them are doing that. But it is completely new to me. I may have some difficulty in following.

B：我知道。但我们从没这样做过，刚开始可能会有点困难。

A：That's fine. Actually, it's quite easy to do.

A：没关系，很容易的。

B：I've never made payment in Renminbi before. It is more convenient for me to pay in dollars.

B：我从来没有用过人民币付款。对我来说用美元付款会更方便些。

A：Many banks in America are in a position to open letters of credit and effect payment in Renminbi against sales confirmation or contract.

A：美国很多银行都办理人民币结算业务，只需要合同。

B: Do you mean to say that I can open a letter of credit in Renminbi with a bank in America?

B：您是说我可以在美国银行开立人民币信用证，对吗？

A: Sure，Consult your banks and you'll see that they are ready to offer you this service.

A：是的，具体业务办理去银行咨询吧。

◇ Dialogue Ⅳ

这段对话发生在杭州红粉服饰有限公司外贸业务员 Dicky 与美国斯特朗服装有限公司采购部经理 Betty 之间，双方讨论修改信用证的问题。

A：杭州红粉服饰有限公司外贸业务员 Dicky

B：美国斯特朗服装有限公司采购部经理 Betty

A: Betty，I need you to inform the bank to amend the letter of credit.

A：Betty，我方需要你方修改信用证。

B: Do you have any questions?

B：有什么问题吗？

A: It is the description of the goods, which should include unit price, quantity and specifications.

A：商品描述方面的问题，商品描述应该把单价、数量、规格等都写清楚。

B: OK, we'll ask our bank to amend the L/C and then fax you the amendment. I believe when you see it again, you will certainly be satisfied with it.

B：好的，我方将通知银行修改信用证，然后用传真将改后的信用证发你方。我方相信你方看到信用证时，一定会满意的。

A: Fine, we'll make an arrangement to ship on receiving the revised L/C.

A：好的，收到改好的信用证后，我方立马安排装运。

B: By the way, I see in the L/C that partial shipment is not allowed. Do you mind making an exception of amending your L/C allowing partial shipment?

B：顺便问一下，我方发现信用证条款规定不允许分批装运。你方介意将信用证改为允许分批装运吗？

A: Sorry, but it's very inconvenient for us. We really must insist on that.

A：对不起，分批装运会给我方带来很大的不便。这一点我方还是坚持的。

◇ Dialogue Ⅴ

这段对话发生在杭州红粉服饰有限公司外贸业务员 Dicky 与美国斯特朗服装有限公司采购部经理 Betty 之间，双方讨论信用证延期的问题。

A：杭州红粉服饰有限公司外贸业务员 Dicky

B：美国斯特朗服装有限公司采购部经理 Betty

A: We have received your letter of credit No.90754 for the amount of US $ 100,000. And the credit calls for shipment on or before April 10.

A：我方已收到你方金额为 10 万美金的第 90754 号信用证。信用证要求在 4 月 10 日前装运。

B: Yes, that's right.

B：是的，没错。

A：We regret to say that, owing to a delay of our suppliers, we will not be able to get the shipment on time. In this case, we have to ask for the extension of L/C and wish you to be so kind as to extend the date of shipment.

A：很抱歉，由于我方供应商的耽误，我们无法按时交货。因此，我们不得不请求延长装运时间并延长信用证的有效期。

B：Would you try your utmost to the original schedule?

B：你方可以尽力在原定时间内交货吗?

A：We hope so, but we have no way to solve it because our suppliers can't offer us goods on time.

A：我方也希望如此，供应商无法按时供货，我方也无能为力。

B：Please tell me the exact date you want to extend L/C.

B：请告知信用证延期的具体时间。

A：May 15th, two-week extension of the validity of the L/C. Is it OK?

A：5 月 15 日，也就是说信用证有效期延长两周。您看行吗?

B：We wish to point out that if you can't get the shipment within the stipulated time, we shall not be able to fulfill our contract with our clients.

B：我方想说明一点，如你方未能按时装运，我方也将无法与客户履约。

A：Only extend 15 days. We will be very appreciative if you can give us an accommodation.

A：仅延长 15 天。如能予我方以通融，我们将不胜感激。

👫 Roles Simulation

Suppose you are one of them in the following conversation. Try to read aloud and practice the underlined sentences.

这段对话发生在杭州红粉服饰有限公司外贸业务员 Dicky 与美国斯特朗服装有限公司采购部经理 Betty 之间，双方讨论使用信用证支付的相关问题。

A：杭州红粉服饰有限公司外贸业务员 Dicky

B：美国斯特朗服装有限公司采购部经理 Betty

A：Shall we talk about the letter of credit again?

A：我们再谈谈信用证的问题好吗?

B：Certainly. Do you have any problem?

B：当然，有什么问题吗?

A：Yes. We have received your L/C No.90754. But after examining, we found that there are three discrepancies between the L/C and the contract.

A：是的。我们已接到你们的第 90754 号信用证。但经审核之后，我们发现信用证与合同之间有 3 个不符点。

B：Really? Let me see.

B：是吗？让我看看。

A：Do you have the L/C at hand?

A：你现在手头上有那份信用证吗？

B：Yes. Go ahead.

B：是的。请讲。

A：Look at your credit, please. Firstly, the Art. No. should read AC- 102 instead of AC-101, the latter being that of your last order.

A：请看信用证。首先，货号应是 AC-102 而不是 AC-101，后者是你方上次订单的货号。

B：Sorry. We made a mistake.

B：抱歉。我们弄错了。

A：And the beneficiary should be written in detail including the full name of the company, the detailed address, the telephone number and the fax number.

A：受益人应该写具体一点，包括公司全称、地址、电话号码、传真等。

B：We are really sorry.

B：我们真是抱歉。

A：Secondly, on perusal of your L/C, we found that the amount isn't very accurate. The correct total comes to $100,000 instead of $10,000. The difference is $90,000.

A：其次，经核查，我们发现信用证金额显然不足。正确总金额应该是 100,000 美元，而不是 10,000 美元，相差 90,000 美元。

B：Oh, sorry. I'm so surprised to hear that.

B：很抱歉。听到这个消息真让我感到意外。

A：There is another point that doesn't conform to the contract: it appears that the quantity in the L/C is not accurate. The correct figure should be 20000 cases.

A：这里还有一个不符点：请将数量修改为：20000 件。

B：We are sorry to trouble you. We will check all the discrepancies at once. If you are right, we will amend it immediately according to your requests.

B：对不起，给你们添麻烦了。我们立刻核对一下不符点。如果真是这样，我们会立即按照你们的要求修改信用证。

A：Besides these. I have another request.

A：另外，我还有一个要求。

B：Yes?

B：哦？

A：Owing to the delay of the establishment of the L/C, we will not be able to ship the goods within the lifetime of L/C which expires on May 10. Because of this case, we have to ask for extension of the validity of the L/C.

A：由于你方开证延误，我方不可能在 5 月 10 日信用证到期日前交货。我方要求延长信用证的有效期。

B：Please tell me the exact date you want to extend L/C.

B：请告知信用证延期的具体时间。

A：May 25th, two-week extension of the validity of the L/C. Is it OK?

A：5 月 25 日，信用证有效期延长两周，您看行吗？

B：Yes, we'll do it according to your request.

B：好的。我们会按照你们的要求进行修改。

A：Thank you very much for your cooperation.

A：谢谢合作。

❤ Typical Expressions

1. You know, we are a small company and opening a L/C is quite costly and will tie up the capital of our company. Can you be a bit more flexible and bend the rules a little? 您知道，我们是一家小公司，开信用证花费很大还占用资金。您能变通一下，换种付款的方式吗？

2. To avoid having our funds tied up, can we make the payment by L/C after sight? 为了避免占用我们资金，可以开远期信用证吗？

3. The L/C must reach us not later than September 27 and remain valid fifteen days after the date of shipment. 信用证应该在 9 月 27 日前到达我方，并在装运日期后 15 天内有效。

4. The shipment time is approaching, but we have not received your L/C. Please speed up the establishment of the relevant L/C, so that shipment may be effected punctually. 发货时间快到了，但我们还没有收到您的信用证。请尽快开立信用证，以便我们及时发货。

5. We have received your L/C, but regret to find that there are some points which do not conform to the contract terms. Please amend them according to the contract. 信用证我们已经收到了。但我们发现有些地方与合同条款不一致，请根据合同要求修改信用证。

6. Our bank advised us today that your transfer of US$1500 was credited to our account. Thank you for paying so promptly, and we hope that you like the consignment and look forward to your next order. 今天银行通知我们 1500 美元已入账。谢谢及时付款，希望你们喜欢这批货物并期待你方下次订货。

7. For small orders, we'd like to make the payment by mail transfer after receipt of the documents. 对于小额订单，我们希望在收到单证后以信汇方式支付。

8. Partial shipment, partial payment. 分批装运，分期付款。

9. I really sympathize with the problem you have had in clearing the balance and am waiting to extend the credit for another six weeks. Would you please confirm that the credit will be settled then? 我对你方结账过程中遇到的麻烦深表同情，愿意展证 6 个星期。请你方确保到时付款好吗？

10. We regret being unable to accept your terms of payment and therefore returning your order. 遗憾很难接受你方付款条件，所以退回订单。

Foreign Trade Letter

◇ **Sample　Paymentarrangement　QC Invoice**

发件人：XXX@XX.com

Date: 2019 年 6 月 15 日　周六下午 6:41

Subject：LF19005 QC INVOICE FOR LIF-08-18-19-2020 and LIF-27-2019S

To: Springway-Roselyne

Cc: Martin

Dear Roselyne,

Hope you had a very nice weekend!

Please kindly find the attached LF19005 QC INVOICE FOR LIF-08-19-2020 and LIf-27-2019S (The amount is $6001.55), we have shipped out goods on 5th June by air for LIF-27-2019S and on 14th June by sea for LIF-08-18-19-2020. Please help to arrange the payment. Thanks!

Best regards

×××

■ **Notes：**

QC：Quality Control　质量控制

Foreign Trade Documents

◇ **Sample　Ⅰ　QC invoice form**

LIK FUNG KNITTING CO.,LTD
QC　INVOICE

TO: SPRINGWAY
Z.A. I. Jollot Curie 3, RUE H. Poincar
93270 SEVRAN, France
SHIPMENT FROM CHINA TO FRANCE

INVOICE NO: LF19005
PAYMENT:
DATE:　15/Jun/19

INVOICE NO	STYLE NO	PO NO	DESCRIPTION	UNIT PRICE Total	QTY (PCS)	UNIT PRICE (FOB CHINA)	TTL AMOUNTS	DEPOSIT
ZS19027S	CRM0287/DM	LIF/27/2019S	LADIES' KNITTED PULLOVER	US$8.80	643	US$5.80	US$3,729.40	US$1,864.70
ZS19008	ZQM0128/HSE-RP	LIF/08/2020	LADIES' KNITTED CARDIGAN	US$13.50	881	US$6.50	US$5,726.50	US$2,863.25
ZS19018	NIZZ/BSH	LIF/18/2020	LADIES' KNITTED CARDIGAN	US$15.80	199	US$6.80	US$1,353.20	US$676.60
ZS19019	NIZZ/BSH	LIF/19/2020	LADIES' KNITTED DRESS	US$18.00	199	US$6.00	US$1,194.00	US$597.00
			THE END					
TOTAL					1922		US$12,003.10	US$6,001.55
TOTOL BALANCE FOR QC INVOICE (DEDUCT 50% DEPOSIT)							US$6,001.55	

◇ **Sample II   Invoice statement**

INVOICE   STATEMENT

CONSIGNEER COMPANY NAME: XXXXX

SHIP BY: DHL

INVOICE NBR: 20200126

SHIP DATE: 2020-1-26

CONIGNEER ADDRESS: XXXXXXX   FRANCE 93270

TEL: :(33)143 XXXX

CONTACT NAME: Chaya

Full description Of Goods(品名描述)	Quantity (数量)	Item Value (单价)	Total Value For Customs (总价)
Ladies 100% acrylic knitted cardigan NO COMMERCIAL VALUE COUNTRY OF ORIGINAL: MADE IN CHINA	8PC	2.0	$16.00

LIFUNG KNITTING FACTORY

SIGNATURE: MARTIN

DATE: 2020-1-26

SAMPLE NOT TO BE SOLD.

ADDRESS: Unit 1758 17/F Da XXX Chamber of Commerce , SouthXXXXX road, XXXX, DongGuan City, Guangdong, China.

TEL: XXXXXXXXX

◇ Sample Ⅲ Letter of credit(draft)

```
{1:F01CCBPFRPPXGRE0000000000}{2:I700ABOCCNBJX170N}{4:
:27:1/1
:40A:IRREVOCABLE
:20:D0682639
:31C:181016
:40E:UCP LATEST VERSION
:31D:190120CHINA
:50:SARL VINTAGE SPIRIT COMPANY
LE TOMAS
HAUTE PLAINE DE CHABOTTES
05260 CHABOTTES TEL 04 92 24 94 05
:59:SUIZHOU LI-FUNG KNITTING CO.,LTD
NO. 1115, Traffic Boulevard,
Zengdu Economic Development Zone,
Suizhou City, Hubei Province, CHINA
:32B:USD42399,3
:39A:3/3
:41D:ANY BANK IN CHINA
BY DEF PAYMENT
:42P:45 days after Bill of lading
:43P:ALLOWED
:43T:NOT ALLOWED
:44A:CHINA
:44E:SHENZHEN PORT, CHINA
:44F:FOS SUR MER PORT , FRANCE
:44B:CHABOTTES, FRANCE
:44C:181230
:45A:FOB SHENZHEN PORT, CHINA
.
AS PER PROFORMA INVOICE DATED 2018-10-10
.
3350 PCS CLOTHES
.
:46A:1) MANUALLY SIGNED COMMERCIAL INVOICE IN 3 FOLDS
.
2) PACKING LIST/ WEIGHT NOTE IN 3 FOLDS  WITH NUMBER OF PACKAGE,
WEIGHT AND SIZE FOR EACH PACKAGE
.
3) FULL SET ORIGINAL CLEAN ON BOARD OCEAN BILL OF LADING
ESTABLISHED TO ORDER OF APPLICANT AS MENTIONNED IN FIELD 50,
NOTIFY APPLICANT IN FULL NAME AND ADRESS AS MENTIONNED
IN FIELD 50,
MARKED FREIGHT COLLECT,
QUOTING OUR L/C,
MENTION ON BOARD,
MUST BE SIGNED BY THE CARRIER OR THE MASTER OR HIS AGENT.
.
4) CERTIFICATE OF ORIGIN ISSUED BY THE SUITABLE OFFICIAL
ORGANIZATION IN THE COUNTRY OF ORIGIN IN ONE ORIGINAL AND ONE
COPY
.
5) COPY OF 1ST SHEET OF QUALITY CONTROL REPORT/INSPECTION REPORT
SIGNED AND ACCEPTED BY THE REPRESENTATIVE OF THE VINTAGE SPIRIT
COMPANY
.
:47A:A) DOCUMENTS MUST BE IN THE LANGUAGE OF THE CREDIT
.
B) 3 PCT MORE OR LESS ON BOTH AMOUNT AND QUANTITY ALLOWED
.
C) 1 P/C PENALTY PER DAY AS FROM 06/01/2019
.
D) IF SHIPMENT IS LATER THAN 13/01/2019
THE SUPPLIER WILL HAVE TO SHIP THE GOODS BY AIR AT HIS EXPENSES.
```

```
IN THIS CASE,
IN FIELD 45A :  CPT MARSEILLE MARIGNANE AIRPORT FRANCE
FIELD 44E : SHENZEN AIRPORT CHINA
FIELD 44F : MARSEILLE MARIGNANE AIRPORT FRANCE
FIELD 46A 3) AIR TRANSPORT DOCUMENT, ORIGINAL FOR SHIPPER,
ISSUED BY AN AIRLINE COMPANY, SHOWING APPLICANT (AS PER FIELD
50) AS CONSIGNEE, NOTIFY APPLICANT AS PER FIELD 50, BEARING
SPECIFIC NOTATION OF FLIGHT NUMBER AND DATE OF SHIPMENT, QUOTING
OUR L/C REF, MARKED FREIGHT PREPAID
.
E) IN ACCORDANCE WITH THE PROVISIONS OF ART 16 C III B OF UCP
600, IF WE GIVE NOTICE OF REFUSAL OF DOCUMENTS PRESENTED UNDER
THIS CREDIT WE SHALL HOWEVER RETAIN THE RIGHT TO ACCEPT A WAIVER
OF DISCREPANCIES FROM THE APPLICANT AND, SUBJECT TO SUCH WAIVER
BEING ACCEPTABLE TO US, TO RELEASE DOCUMENTS AGAINST THAT WAIVER
WITHOUT REFERENCE TO THE PRESENTER PROVIDED THAT NO WRITTEN
INSTRUCTIONS TO THE CONTRARY HAVE BEEN RECEIVED BY US FROM THE
PRESENTER BEFORE THE RELEASE OF THE DOCUMENTS.
ANY SUCH RELEASE PRIOR TO RECEIPT OF CONTRARY INSTRUCTIONS SHALL
NOT CONSTITUTE A FAILURE ON OUR PART TO HOLD THE DOCUMENTS AT
THE PRESENTER'S RISKS AND DISPOSAL, AND WE WILL HAVE NO
LIABILITY TO THE PRESENTER IN RESPECT OF ANY SUCH RELEASE.
.
F) IN CASE THE L/C IS AMENDED THE BENEFICIARY WILL HAVE TO
PRODUCE A CERTIFICATE STATING THAT THE AMENDEMENTS HAVE BEEN
ACCEPTED AND PRECISING L/C REFERENCE, THE NUMBER OF THE
AMENDEMENTS AND THEIR ISSUANCE DATE.
IN CASE THE L/C IS NOT AMENDED SUCH CERTIFICATE IS NOT REQUIRED.
.
G) IF THE PRESENTED SHIPPING DOCUMENTS INCLUDE ANY REFERENCE TO
COUNTRIES,REGIONS, ENTITIES, VESSELS OR INDIVIDUALS SUBJECT TO
ANY APPLICABLE INTERNATIONAL SANCTIONS REGIMES AND RELEVANT
REGULATIONS IMPOSED BY US/EU/UN, WE SHALL NOT BE LIABLE FOR ANY
DELAY OR FAILURE TO PAY, PROCESS OR RETURN SUCH DOCUMENTS.
.
:71B:BANKING CHARGES INCURED OUTSIDE
FRANCE AND/OR DISCOUNT CHARGES
FOR BENEF'S ACCOUNT. WE WILL
DEDUCT A DISCREPANCY FEE OF EUR
140.00 OR EQUIVALENT FOR DRAWING
WHICH DOES NOT COMPLY WITH L/C.
:48:DOCUMENTS TO BE PRESENTED AT
NEGOTIATING BANK WITHIN 21 DAYS
AFTER SHIPMENT DATE.
:49:WITHOUT
:78:AT THE BEST CONVENIANCE OF THE PRESENTING/NEGOTIATING BANK ON
MATURITY DATE IF DOCUMENTS COMPLY STRICTLY WITH CREDIT TERMS AND
CONDITIONS.
KINDLY ACKNOWLEDGE RECEIPT.
PRESENTING/NEGOTIATING BANK MUST CONFIRM THAT THE AMOUNT OF EACH
DRAWING HAS BEEN NOTED ON THE REVERSE OF THIS CREDIT.
WE SPECIFY WE ARE AFFILIATED WITH NATIXIS PARIS (NATXFRPP).
.
AN EXTRA COPY OF ALL DOCUMENTS IS REQUIRED FOR ISSUING BANK'S
FILE, OTHERWISE A PHOTOCOPY HANDLING FEE OF EUR20.00 OR
EQUIVALENT PER SET WILL BE DEDUCTED FROM PAYMENT.
:57D:AGRICULTURAL BANK OF CHINA HUBEI
JINJIN PARK NO.66 ZHONGBAI ROAD
WUHAN CITY HUBEI PROVINCE
ACC 17-783401040001466 ABOCCNBJ170
:72:DOCTS TO SEND IN 1 LOT BY ANY FAST
COURIER SERVICE TO :
BANQUE POPULAIRE AUVERGNE RHONE
ALPES,
```

Page 3 sur 3

```
2 AVENUE DU GRESIVAUDAN,
38700 CORENC(FRANCE).
-}
```

◇ Sample IV   Commercial invoice

COMMERCIAL INVOICE

1. Shipper/Exporter			8. Invoice No.and date			
YOKI TRADING CO.,LIMITED			YK120701	2012-6-31		
UNIT 04, 7/F, BRIGHT WAY TOWER, NO. 33			9. NO. & DATE OF L/C			
MONG KOK ROAD, KOWLOON, HK.			NO L/C			
Tel: 00852-23892981	FAX:00852-35902333					
2. For Account & Risk Of Messers						
SALT AND PEPPER CLOTHING, INC.						
TEL : 323-232-XXX FAX:323-232-XXX						
3. Notify Party			11. REMARK			
SALT AND PEPPER CLOTHING, INC.			* Consignee:			
4770 E.50TH ST VERNON, CA 90058						
TEL : 323-232-XXXX FAX:323-232-XXXX			SALT&PEPPER			
4. Port of Lolding GUANGZHOU,CHINA	5.Final Destintation					
6. Carrier	LOS ANGELES, CA, USA					
BY SEA	7. Sailing on or About					
12.Marks and Nos. of PKGS	8/25/2012		14. Quantity	15. UNIT-PRICE	16. Amount	
10 CTN	13. Description of Goods WOMEN DRESS		986 PCS	$6.90	$6,803.40	
HANDLE WITH CARE				FOB GUANG ZHOU, CHINA PORT		
SALT & PEPPER	PO NO	STYLE NO	ITEAM	Quantity	Unit price	AMOUNT
	SNP86506	JD1759		986PCS	US$6.90	US$6,803.40
CARTON						
P.O NUMBER(purchase order)						
STYLE NUMBER						
COLOR						
SIZE						
COUNT						
TOTAL PACK						
TOTAL Q'TY						
MADE IN CHINA						
	TOTAL			986PCS		US$6,803.40
			17. SINEND BY			

◇ Sample Ⅴ Country of original (CO)

ORGINAL					
1.Exporter		Certicate No.			
HUIZHOU BOYUANXIN INDUSTRIAL CO., LTD					
ARE BEHALF OF ZONGSEN TEXTILE CO.,LIMITED					
2.Consignee		CERTIFICATE OF ORIGIN OF THE PEOPLE'S REPUBLIC OF CHINA			
FOREVER21 INC					
3880 N.MISSION ROAD					
DOCK H 18					
LOS ANGELES,CA 90031					
3.Means of transport and route		5.For certifying authority use only			
FROM SHANGHAI TO LOS ANGELES USA BY SEA					
4.Country/regin of destination					
UNITED STATES					
6.Marks and numbers	7.Number and kind of packages;description of go	8.H.S.Code	9.Quantity	10.Number and date of invoices
CARTON				
P.O NUMBER	PO# 15175426	6204440090	953PCS	ZS15140
STYLE NUMBER	DRESS			12-Jan-16
COLOR	TOTAL: EIGHTY ONE(81)CTNS ONLY			
SIZE				
COUNT				
TOTAL PACK				
TOTAL Q'TY				
MADE IN CHINA				
11.Dectaration by the exporter		12.Certification		
The undersigned hereby declares that the above details and statem are correct,that all the goods were produced in China and that they comply with the Rutes of Origin of the People's Republic of China.		It is hereby certified that declaration by the exporter is correct		
Pace and date,signature and stamp of authorized signatory		Pace and date,signature and stamp of authorized signatory		

◇ Sample Ⅵ SGS inspection report

SGS

SGS-P-7.4-01-F02-CRS-SL

Report No.: DGGWT00021416

SGS-CSTC Standards Technical Services Co., Ltd. Dongguan Branch
8/F, Haoyu Building No. 309, Qifeng Road, Guan Cheng District, Dongguan, China P.C:523129
Tel:86 769 22320568
Fax:86 769 22320580/22320060

INSPECTION REPORT

To :	SARL VINTAGE SPIRIT COMPANY	Date:	19-Jul-2018
Attn :	HELEN		
From :	Molly Mo	E-mail :	molly.mo@sgs.com

SGS File No.:	CNSZX19206728	Product family view
Buyer :	SARL VINTAGE SPIRIT COMPANY	
Supplier :	ZONGSEN KNITTING FACTORY	
Manufacturer :	LIFUNG-FASHION	
Style Number:	PH8KN11	
Product description:	MEN'S CARDIGAN	
P.O. Number:	VSC-POFW18-01ZONGSEN	
L/C Number:	LC18050000403737	
Service performed :	FRI	
Inspection Date :	18-Jul-2018	
Inspection Location :	Dongguan, Guangdong, China	

Inspection Criteria

Reference sample provided by	By Manufacturer
Client instruction/specification	YES
SGS WI number	P-INSP-WI-SL-001
Other	N/A

Overall Inspection Conclusion:	**Not Conform**

Inspection Summary:

1. Workmanship appearance :	Conform
2. Quantity :	Conform
3. Style, Material, Colour :	Conform
4. Data measurement / Function/Field tests:	Not Conform
5. Packing :	Subject to client's evaluation
6. Marking / Label :	Conform

Problem Remark:

Found the extra button was put into a transparent bag then attached on care/composition label, but spec required that add the extra buttons on the care label.

Inspector: Guohui Lin   Factory Representative: Ms. Xu

Oral Practice

Forming a group of two or three partners. Try to work out a dialogue related to the theme of this unit and then perform it in class.

Self-Assessment

Evaluate your practice by marking the following each corresponding item and work out the total scores.

考核情境(分值)	考核要求(分值)	得分
语音语调(20分)	发音准确，声音清晰(10分)	
	语调自然，语速自然流畅(10分)	
内容(30分)	内容完整(10分)	
	内容表述符合给定情境(10分)	
	前后内容逻辑一致，内容之间无矛盾(10分)	
语言描述(30分)	用词规范、准确，无语法错误(20分)	
	表述无歧义(10分)	
其他(20分)	无雷同状况(10分)	
	无严重偏题(10分)	
合　计(100分)		

Task 8   Negotiation on Contract

知识目标

◆ 学习国际贸易中合同谈判的内容
◆ 学习国际贸易中合同谈判的艺术
◆ 学习相关商务沟通的口头英文表达方式
◆ 学习相关商务沟通的书面英文表达技巧

能力目标

◆ 能够准备合同谈判材料
◆ 能够顺利开展合同谈判活动
◆ 能够使用双语顺畅地进行商务口头沟通
◆ 能够处理合同谈判往来信函

Basic Knowledge Review

1. What is contract negotiation?

Contract negotiation is the process through which two or more parties come to a legally binding agreement on the terms of their relationship. The main goal of contract negotiation is for each party to be satisfied with the rights and obligations assigned to them, and ready to sign.

2. How do we achieve the goal of negotiation on contract successfully?

The following tips can help you to achieve the goal of negotiation on contract in the international trade.

(1) Create a road map for your export journey

Your international agreement should clarify all the details related to working with the overseas parties related to your business. Comprehensive agreements should clearly define the rights and responsibilities of each party to the agreement. Ideally, your contract should also include all intermediaries in the supply chain.

(2) Spell out expectations and sales goals

A key component of every contract with an overseas representative is accountability. Performance expectations should be clearly spelled out, along with the requirements for satisfactory performance related to volume or defined time periods.

Exclusivity is another important decision. Any exclusivity rights should always be well-defined, and identify exactly what the exclusivity will and will not apply to.

(3) Define the dispute resolution process

The dispute resolution section of a contract is an important element and should clearly define the rights and remedies of all parties to the agreement. This section is particularly important when it comes to facilitating your ability to potentially recover damages in the event of a dispute between one or more of the parties to an agreement.

(4) Include an arbitration clause

One of the cornerstones of a well-drafted dispute resolution section is that it specifies how disputes will be resolved. For exporters, an international arbitration clause can help level the playing field in the event of a dispute.

Failure to include an international arbitration requirement may potentially force your company to choose between litigating claims in a foreign jurisdiction, under its foreign laws, or walking away with the loss. If the prospective parties to your contract are not willing to agree to an international arbitration clause, you have to ask yourself what level of business risk you are willing to take.

(5) Define the legal jurisdiction

The dispute resolution section of your contract also should clearly articulate the controlling law governing the agreement, including the specific legal jurisdiction.

Another important consideration is the actual language used to draft the contract. Contracts used by U.S. importers are often written in English and the official language of the country in which the agreement covers. But it is important to keep in mind that the local language version may create disputes. For that reason, any agreement should be consummated with certified translations.

Also, if your contract is negotiated by a U.S.-based attorney, most likely he or she will not be licensed to practice law in any of the foreign jurisdictions involved in the agreement. In some cases, you may need to retain counsel in the foreign jurisdictions in order to coordinate with your company's primary attorney.

(6) Anticipate contract termination

Your contract should also specifically outline the steps necessary for a valid termination, without a material breach of the agreement. For example, how far in advance must your partner be notified, and how should that notification be delivered? Also, it is important to define what situations might be grounds for termination, particularly before the end of a defined contract term.

Task Descriptions

杭州红粉服饰有限公司代表与美国斯特朗服装有限公司代表就价格、运输、保险与支付谈判成功之后，开始进入产品合同相关问题的谈判。

Negotiation on Contract

接下来的对话就是杭州红粉服饰有限公司代表与美国斯特朗服装有限公司代表围绕产品合同谈判等一系列活动展开的。

Typical Dialogues

◇ Dialogue Ⅰ

这段对话发生在杭州红粉服饰有限公司外贸业务员 Dicky 与美国斯特朗服装有限公司采购部经理 Betty 之间，双方就合同细节进行再确认。

A：杭州红粉服饰有限公司外贸业务员 Dicky

B：美国斯特朗服装有限公司采购部经理 Betty

A：Well，I have brought with me the draft of our contract. Please have a look and let us know anything you are not clear about.

A：合同的草稿我已经带来了。请您看一下还有没有不清楚的事。

B：Well, there is something we should add to this provision: "If one side fails to honor this contract, the other side is entitled to cancel this contract." Do you think so?

B：好的。应该再增加一条："如果一方不履行合同，另一方有权取消合同。"您看呢？

A：Certainly. Anything else you've noticed?

A：可以。还有吗？

B：There is one more thing to make clear. I'd like to know in case of claims, which institution in China will handle arbitration.

B：还有一件事需要弄清楚。万一发生纠纷，中方仲裁机关是哪个？

A：Oh, suppose we have a dispute, we can resolve the case by submitting the dispute to the China International Economic and Trade Arbitration Commission.

A：哦，假如我们有争议，我们可以提交中国国际贸易仲裁委员会解决。

◇ Dialogue Ⅱ

这段对话发生在杭州红粉服饰有限公司外贸业务员 Dicky 与美国斯特朗服装有限公司采购部经理 Betty 之间，双方共同修改合同。

A：杭州红粉服饰有限公司外贸业务员 Dicky

B：美国斯特朗服装有限公司采购部经理 Betty

A：Is the contract all right now?

A：合同这样可以了吗？

B：Um-hmm, We can't sign the contract yet. There are still some changes to be made. This clause in the contract, for example, will have to be changed.

B：嗯，我们还不能在合同上签字。有些地方还得修改。比如说，合同的这一条，就必须改一下。

A：Would you like to do that right now?

A：现在就改吗？

B：Now is as good a time as any. Let me do that.

B：现在就改，我来改。

A：Highlight it with a red pen, we will retype it.

A：修改的地方请用红笔标注，我们会重打。

B：Done! Can you read that?

B：改好了。看得清楚吗？

A：Oh, sure.

A：哦，没问题。

B：We'll have to get this contract done as soon as possible. Send the final contract to me by EMS after signing it.

B：我们得尽快把这份合同做出来。合同签字后尽快快递发我方签字。

A：No problem.

A：没问题。

B：Wait a moment, I think we need to discuss the fifth clause again and I need to talk to the home office.

B：等一下，我想第五条还需要再讨论一下，我还要先跟总公司联系一下。

A：Fine. Just let us know when you are ready. By the way, who is going to sign the contract for your side?

A：好吧，准备好了就请通知我们。对了，谁代表你们这一方签约？

B：The general manager.

B：我们公司总经理。

◇ Dialogue Ⅲ

这段对话发生在杭州红粉服饰有限公司外贸业务员 Dicky 与美国斯特朗服装有限公司采购部经理 Betty 之间，双方就合同中个别条款进行再确认。

A：美国斯特朗服装有限公司采购部经理 Betty

B：杭州红粉服饰有限公司外贸业务员 Dicky

A：Well, it seems that we have talked about almost everything.

A：呃，看起来我们已经谈得差不多了。

B：Yes. I'm glad our discussion has come to a successful conclusion. Are we anywhere near a contract yet?

B：是的，很高兴我们的磋商能成功。我们是不是要准备签合同了？

A：There is a particular clause in our contract I'd like to discuss with you.

A：我还想就合同中的个别条款和您讨论一下。

B：I know it is the disputes settlement mechanism.

B：我知道是争端解决机制。

A：That's right. What would you do in case of disputes? Can you tell me something about that?

A：是的。万一有争议的话，你们是怎么解决的呢？能给我说说你们的做法吗？

B：Yes, of course. If there should be any disputes, we always follow the principles of independence and initiative, equality and mutual benefit. We try to have them settled through friendly negotiation and conciliation.

B：当然可以。如果我们之间出现争议，我们是按独立自主、平等互利的原则来办事的。我们努力通过友好协商和调解来解决。

A：Sounds fair to me. But I still want to have a separate clause for it, OK?

A：我看这很公平合理。但我还是希望将此条款单独列出，可以吗？

B：No problem. It can be written in the contract as "In case of a dispute, it shall be first settled through friendly negotiation, and if unsettled, it shall be submitted to arbitration by China International Economic and Trade Arbitration Commission". Do you agree?

B：没问题。我们可以在合同里写上"如有争议，首先应通过友好协商解决，如果协商不能解决，就提交中国国际贸易仲裁委员会去仲裁。"您同意吗？

A：Good. I hope we'll never invoke this particular clause.

A：好。希望我们永远不会有机会使用这一条款。

◇ **Dialogue Ⅳ**

这段对话发生在杭州红粉服饰有限公司外贸业务员 Dicky 与美国斯特朗服装有限公司采购部经理 Betty 之间。Dicky 将准备好的合同与 Betty 在签合同前就合同文本进行最后的确认。

A：美国斯特朗服装有限公司采购部经理 Betty

B：杭州红粉服饰有限公司外贸业务员 Dicky

A：I think we need to take the copies of our contract into consideration. How many originals of the contract have you prepared?

A：我想我们应该考虑一下合同文本的事情。您准备了几份合同原件呢？

B：There are two originals of the contract both in Chinese and English.

B：两份。每份都有中英文两个版本。

A：They're equally authentic in terms of law.

A：两份都具有同等法律效力。

B：Yes. Here's a copy for you to check.

B：是的，您看一下吧。

A：Thank you. Very good！When can the contract be ready for signing?

A：谢谢！很好，合同什么时候能签呢？

B：I will have it ready in two days.

B：过两天吧。

A：Good! I'd like to look over the contract before I sign it.

A：好。签约之前我想再核对一遍。

B：Of course. I'll take it over for your perusal when it gets ready.

B：当然，合同做好后一定会再给您过目的。

A：Fine，thanks.

A：好的，谢谢。

◇ **Dialogue Ⅴ**

这段对话发生在杭州红粉服饰有限公司外贸业务员 Dicky 与美国斯特朗服装有限公司采购部经理 Betty 之间，合同签订完毕之时。

A：杭州红粉服饰有限公司外贸业务员 Dicky

B：美国斯特朗服装有限公司采购部经理 Betty

A：We've finally come to this moment.

A：终于等到这一刻了。

B：Yes, I've been looking forward to it, too.

B：是的，我也一直期待着这一刻。

A：Each of us will sign two formal copies of the contract, the Chinese copy and the English one. Here they are. Please sign it here and here.

A：正式合同一式两份，英文和中文各一份。请在这里和这里签字。

B：(After signing) We've made it at last. Allow me to propose a toast to the success of our business and to our future cooperation. Cheers!

B：(签字后)终于签合同了。我提议为我们的合作成功干杯！

A：Cheers!

A：干杯！

👥 Roles Simulation

Suppose you are one of them in the following conversation. Try to read aloud and practice the underlined sentences.

这段对话发生在杭州红粉服饰有限公司外贸业务员 Dicky 与美国斯特朗服装有限公司采购部经理 Betty 之间，双方就合同相关问题进行最后的确定。

A：杭州红粉服饰有限公司外贸业务员 Dicky

B：美国斯特朗服装有限公司采购部经理 Betty

A：Here is the contract. They are two originals of the standard contract we prepared. It contains basically all we have agreed upon during our negotiation.

A：这是合同。两份合同原件是我们按照合同范本做的，内容涵盖谈判期间达成的所有协议。

B：Don't you think it necessary to have a close study of the contract to avoid anything missing?

B：您觉得有没有必要再仔细看一下，还有没有遗漏的地方？

A：Well, then, we've agreed on all the major points. I'd like to say something about the copies of our contract.

A：嗯，合同的主要内容我们已经达成一致意见了。我想先说说合同文本的问题。

B：How many originals of the contract have you prepared?

B：您已经准备了多少份合同的原件？

A：There are two originals of the contract both in Chinese and English.

A：合同正本一式两份，每份都有中英文两个版本。

B：They are equally authentic in terms of law.

B：两份合同在法律上都真实有效。

A：Yes. Here's a copy for you to check.

A：是的。这份副本供您核对。

B：Generally speaking, a contract cannot be changed after both parties have signed it.

B：一般来说，合同经双方签字后不能变更。

A：So we'd better make sure one more time that we've got them right?

A：所以我们最好再核对一遍。

B：To make sure no important items have been overlooked, shall we check all the terms and conditions of the transaction to see if there is anything not in conformity with the terms we agreed on?

B：为了确保没有重要的项目被忽视，我们再检查一下交易的所有条款，看看有没有我们不同意的条款，好吗？

A：Okay, let's start from the name of the commodity, specification, quantity, unit price….

A：好吧，让我们从商品的名称、规格、数量、单价开始。

B：All right. We have no objection to the stipulations about all the terms and conditions.

B：好吧。合同所有条款我方都认同。

A：We have no objection to the stipulations about all the terms and conditions, too.

A：合同所有条款我方也都认同。

B：Now, are we ready to sign?

B：现在我们双方可以签字了吗？

A：Yes, I've been looking forward to this moment.

A：是的，期待这一刻很久了。

B：I'll send the copies of the contract to you by EMS after we sign the contract. Then you should send one copy of the contract back to me after you countersign it.

B：我方签完字后，将用快递将合同送达你方。你方回签后，再将合同寄回一份。

A：That is fine，Thank you. I'm glad the deal has come off nicely and hope there will be more to come.

A：好的，谢谢你。很高兴一切都很顺利，并希望将来有更多的合作。

B：So long as we keep to the principle of equality and mutual benefit, trade between our countries will develop further.

B：只要我们坚持平等互利的原则，我们两国之间的贸易就会进一步发展。

A：Let's congratulate ourselves on having brought this transaction to a successful conclusion.

A：为我们合作成功庆祝吧!

B：Congratulations!

B：恭喜!

❧ *Typical Expressions*

1. I still have some questions concerning our contract. 关于合同内容，我还有点疑问。

2. The negotiation on rights and obligations of the parties under contract turned out to be very successful. 关于合同双方权利义务的谈判非常成功。

3. The contract will be sent to you by air mail for your signature. 合同将通过航空邮件发你方签字。

4. Don't you think it necessary to have a close study of the contract to avoid anything missing? 你方难道不想将合同再仔细捋一遍，看看有没有需要补充的内容吗?

5. We'll ship our goods in accordance with the terms of the contract. 我方将按照合同条款装运货物。

6. The dresses will be made of the best materials and the stipulations of the contract will be strictly observed. 这批衣服严格按合同的规定采用上好的料子制作。

7. If there is a changing situation at one end and the work cannot be completed at the time originally stipulated in the contract, they have to request the other party to amend relative contract terms. 如一方情况有变，不能按合同规定时间完成的，必须要求另一方修改相关合同条款。

8. No party should amend the contract unilaterally without the other party's agreement in writing. 未经对方书面同意，任何一方不得单方面修改合同。

9. We have every reason to cancel the contract because you've failed to fulfil your part of it. 我方完全有理由取消合同，因为你方未履行合同。

10. Generally speaking, a contract cannot be changed after it has been signed by both parties. 一般来说，合同在双方签字后就不得更改。

✉ *Foreign Trade Letter*

◇ **Sample   Order confirmation**

DEAR SIMI,

THANKS TO FIND ENCLOSED THE ORDER LIF/17/2020-GLM0386/CLA

-PRICE: THANKS TO SETTLE PRICE WITH MAURICE DURING MEETING.

-SHIPMENT: THANKS TO CFM 28/05 ETD VESSEL.

THANKS TO SEND URGENTLY THE SAMPLES REQUESTED BY CHAYA ON 14/02.

Sincères salutations/Best rgds

Spring way

Foreign Trade Documents

◇ Sample Ⅰ   Purchase order sample

SPRINGWAY

Z.A.I. Joliot Cur FOLLOWED BY THI-OANH

3, RUE H. Poinc. SPRINGWAY QC IS REQUIRED

2019/7/6

93270 SEVRAN

France

OUR CONTRACT	LIF/10/2020
OUR STYLE N°	SU-H19-16-LIF/SU
BUYER ARTICLE N	270001

SUPPLIER: LI-FUNG

Traders Compai   CONTACT: MARTIN/SIMI

BUYER ORDER

DESCRIPTION:				
	LADIES PULLOVER - ROUND NECK - LONG SLEEVES			
PRICE	PRICE TBC with MAURICE ROSANEL PRICE ... USD ... / FOB SHZ (including basically labelling)		CONTENT:	TBC
SHIPMENT:			STYLE WITH TONE/TONE SATIN HANGER LOOPS	
	21/05/2019 – ETD VESSEL			
SHIPPING MARK:		SIDE MARK:		
QUANTITY:	2540 PCS			
COLOURS:				
		GROSS Wt:		
SIZES:	36 TO 46/48	NET Wt:		
CONTRACT N°:	LIF/19/2020			
STYLE:	SU-H19-16-LIF/SU			

QUANTITY:	2540	PIECES

ACCESSORIES AND LABELLING TO BE PROVIDED BY SELVY ASAP

◇ **Sample Ⅱ Sales confirmation**

LIK FUNG KNITTING COMPANY LIMITED

SALES CONFIRMATION

DATE : 30-Oct-20

REF. NO. : SC131101

BUYER: SPRINGWAY S.A,   SELLER : LIK FUNG KNITTING COMPANY LIMITED

STYLE	YARN CONTENT	QUANTIY	UNIT PRICE(RMB)	AMOUNT	DEPOSIT	BALANCE
D0934/DM	2/45NM 54% POLYESTER 20% ACRYLIC 20% NYLON6% MERINO WOOL	6000	￥58.00	￥348,000.00	￥104,400.00	￥243,600.00
TOTAL		6000		￥348,000.00	￥104,400.00	￥243,600.00

SAY RMB TWENTY THOUSAND SEVEN HUNDRED SIXTY FOUR ONLY

DELIVERY SCHEDULE: 20,Dec,2013

SHIPMENT TERM : FOB SHENZHEN

PAYMENT : 30% DEPOSIT IN 7 DAYS. BALANCE BY T/T BEFORE SHIPMENT.

THE BUYER :
SPRINGWAY

THE SELLER :
LIK FUNG KNITTING COMPANY LIMITED

AUTHORIZED SIGNATURE   AUTHORIZED SIGNATURE

◇ Sample Ⅲ   DDP payment purchase order

FOREVER 21, Inc.

BILL TO: FOREVER21, INC. 3880 N. Mission Road Los Angeles, CA 90031 attn:mport-ap@forever21.com	SHIP TO: FOREVER 21, INC. 3880 N. Mission Road Los Angeles, CA 90031 PHONE : 213-741-8866	Order Date: 07/08/2013	CO: China	ITEM CODE: 31558198

Order Date: 07/08/2013 CO: China

EX C.O Date: 09/30/2013   LABEL: F21MAINB03

In House Date: 10/21/2013   Price Ticket: Y   Sticker: N

VENDOR: 15667

Pay Type: WIRE DDP   Hanger Tape: N   Hang Tag:

Ship Via: OCEAN   Terms: OTHER

Material	Pattern	Season
SWEATER	ANIMAL	

Ship By:
GARMENT ON HANGER   N
CONFORM TO F21PACKINGINSTRUCTION   Y

VALERIE CHAO/MAIN

PO#	STYLE#	COST	DESCRIPTION	REMARKS	TREND	FS
13047135	MKC1306211 (USA)	9.90	SWEATER TOP/LSLV	NEW, XXI,F21,CAN,MEX,COL,WEB		FASH

BUYER* LEGEND	USA=FOREVER21,INC	JPN=FOREVER21JAPAN RETAIL	CAN=FOREVER XXI,ULC	KOR=FOREVER21 KOREA RETAIL	EU=FOREVER21 GLOBAL B.V	PHL=FOREVER AGAPE & GLORY
CH=FOREVER 21 COMMERCIAL(SHANGHAI) LIMITED		HK=FOREVER21ASIA HOLDINGS LIMITED		ISR=FOREVER21ISRAEL LTD.	MEX=FOREVER 21 MEXICO, S. DE R.L. DE C.V.	COL=ALAMEDA COLOMBIA S.A.S.
CRI=SISAL CR S.A	IND=DIANA RETAIL PRIVATE LIMITED	PAN=ALAMEDA PANAMA S.A	CHL=SISAL CRILE SPA	SLV=SISAL ES,S.A.DE C.V.		

CID	COLOR	*BUYER	LABEL	P/T	XS	S	M	L	XL	1X	2X	3X	4X	TOTAL
02	HEATHER GREY/BLACK	USA	F21	USA	0	2,591	1,944	1,300	0	0	0	0	0	0	5,835
		USA	XXI	USA	0	80	60	40	0	0	0	0	0	0	180
		CAN	F21	CAN	0	440	330	220	0	0	0	0	0	0	990
		MEX	F21	MEX	0	52	39	26	0	0	0	0	0	0	117
		COL	F21	COL	0	108	81	54	0	0	0	0	0	0	243
		USA-W	F21	USA	0	576	432	288	0	0	0	0	0	0	1,296
				TOTAL	0	3,847	2,886	1,928	0	0	0	0	0	0	8,661
				TOTAL											8,661

RELATED P.O#	SHIP ADDRESS	IN HOUSE	QTY
13047140	BONDED (KOREA, JA	09/30/2013	1,733
13047143	NON-BONDED:CHINA	09/30/2013	210
	ITEM TOTAL		10,604

IMPORTANT: THE TERMS AND CONDITIONS ON THE REVERSE SIDE OF THIS PURCHASE ORDER ARE INCORPORATED HEREIN AND EFFECT YOUR LEGAL RIGHTS. PLEASE READ AND REVIEW BOTH CAREFULLY BEFORE SIGNING THIS AGREEMENT.THIS ORDER IS MARKED SAMPLE TO BE APPROVED.

VENDOR SIGNATURE: _____   VENDOR NAME: _____

THE LAW OF THE STATE OF CALIFORNIA INCLUDING THE UNIFORM COMMERCIAL CODES AS ENACTED BY THE STATE OF CALIFORNIA GOVERNS ALL SALES TO FOREVER 21 MERCHANDISING UNDER THE P.O.

ANY DEVIATION TO ABOVE INSTRUCTIONS WILL RESULT IN A CHARGE BACK

ABC Oral Practice

Forming a group of two or three partners. Try to work out a dialogue related to the theme of this unit and then perform it in class.

Self-Assessment

Evaluate your practice by marking the following each corresponding item and work out the total scores.

考核情境(分值)	考核要求(分值)	得分
语音语调(20 分)	发音准确，声音清晰(10 分)	
	语调自然，语速自然流畅(10 分)	
内容(30 分)	内容完整(10 分)	
	内容表述符合给定情境(10 分)	
	前后内容逻辑一致，内容之间无矛盾(10 分)	
语言描述(30 分)	用词规范、准确，无语法错误(20 分)	
	表述无歧义(10 分)	
其他(20 分)	无雷同状况(10 分)	
	无严重偏题(10 分)	
合　计(100 分)		

Task 9　Negotiation on Complaints

知识目标

◆ 学习国际贸易中争议谈判的内容
◆ 学习国际贸易中争议谈判的艺术
◆ 学习相关商务沟通的口头英文表达方式
◆ 学习相关商务沟通的书面英文表达技巧

能力目标

◆ 能够准备争议谈判材料
◆ 能够顺利开展争议谈判活动
◆ 能够使用双语顺畅地进行商务口头沟通
◆ 能够处理争议谈判往来信函

Basic Knowledge Review

1. What does negotiation on complaints involve?

The Dispute Settlement Understanding (DSU) is the main WTO agreement on settling disputes.

In summary, the WTO dispute settlement system provides for three kinds of complaints: "violation complaints", "non-violation complaints" and "situation complaints". Violation complaints are by far the most frequent.

Generally speaking, the following type of complaints maybe considered:

(1) Complaints received from foreign buyers in respect of poor quality of the products supplied by exporters.

(2) Complaints of unethical commercial dealings categorized mainly as non-supply/partial supply of goods after confirmation of order; supplying goods other than the ones as agreed upon; non-payment' non-adherence to delivery schedules, etc.

2. How do we achieve the goal of negotiation on complaints successfully?

Here are checklist for negotiation on complaints to help you settle the differences successfully.

(1) Initial assessment:

✓   Authority/Mandate to negotiate and reach an agreement or settlement
✓   Willingness to negotiate
✓   Credibility of other party(ies)
✓   Ability to negotiate (equality?)
✓   Alternatives to negotiation

(2) Contact with the other party to arrange/confirm:
✓   Agenda
✓   Location (neutral)
✓   Timetable
✓   Participating parties
✓   Public/Confidential nature (See statutory requirements, below)
✓   Official Languages
✓   Support services (word processing, etc.)

(3) Preparation of a strategy and interest assessment:
✓   Study the issues
✓   Harmonize/reconcile competing interests within the team
✓   Assess the BATNA (Best Alternative to a Negotiated Agreement) for all parties
✓   Assign roles for team members (spokesperson(s), etc.)
✓   Create options for mutual gain ("win-win")
✓   Consult relevant statutes (including the Access to Information Act, the Department of Justice
Act, the Official Languages Act, the Privacy Act) and relevant policy directives

(4) Pointers for a negotiation:
✓   Concentrate on interests, not positions
✓   Separate the people from the problem
✓   Listen carefully and actively
✓   Respect the other party (e.g., any cultural, linguistic or other differences)
✓   Create and propose options for mutual benefit ("win-win")
✓   Use objective standards
✓   Assess progress in light of one's BATNA
✓   Caucus if necessary
✓   Anticipate and avoid responding to provocative tactics
✓   Communicate frequently with the client
✓   Remain within the limits of the negotiating mandate

✒ Task Descriptions

Negotiation on Complaints

杭州红粉服饰有限公司代表与美国斯特朗服装有限公司代表签订合同后，双方按合同要求履行各自的义务。但是，双方在执行合同过程中发生种种意外，致使产品、运输、保险与支付等方面不能履约。

接下来的对话就是杭州红粉服饰有限公司代表与美国斯特朗服装有限公司代表围绕合同履行过程中发生的争议等一系列谈判活动展开的。

♬ Typical Dialogues

◇ Dialogue Ⅰ

这段对话发生在杭州红粉服饰有限公司外贸业务员 Dicky 与美国斯特朗服装有限公司采购部经理 Betty 之间，双方讨论货物质量与原合同不一致的问题。

A：美国斯特朗服装有限公司采购部经理 Betty

B：杭州红粉服饰有限公司外贸业务员 Dicky

A：Hi, Dicky. I am calling about a mistake on the last order in October. I am afraid I'll have to cancel our order.

A：Dicky，我打电话想说一下 10 月份订单出事了。恐怕我将不得不取消我们的订单。

B：What is it?

B：怎么了？

A：The dresses you delivered do not match the final sample you provided. We cannot accept it.

A：你方发来的货物与你方提供的确认样不符。我方不能接受。

B：I'll look into this matter immediately and give you a satisfactory reply.

B：我将立即调查此事，并给您一个令人满意的答复。

A：If you don't reduce your price, I'll consider to cancel the deal.

A：如果你方不降价，我方会考虑取消这笔交易。

B：What kind of price did you have in mind?

B：你方觉得多少价格合适？

A：50% off, do you agree?

A：一半可以吗？

B：Well, in view of our long good relationship, we can make major concessions for this order.

B：好吧，鉴于多年友好合作关系，这批衣服我们可以让步。

A：Thank you. Let's call it a deal.

A：谢谢。那么成交了。

B：All right.

B：好吧！就这么定了。

◇ Dialogue Ⅱ

这段对话发生在杭州红粉服饰有限公司外贸业务员 Dicky 与美国斯特朗服装有限公司采购部经理 Betty 之间，双方讨论货物数量与原合同不一致的问题。

A：美国斯特朗服装有限公司采购部经理 Betty

B：杭州红粉服饰有限公司外贸业务员 Dicky

A：Hi, Dicky. It seems that there was something wrong with the shipment.

A：Dicky, 你好，货有问题。

B：What is the matter?

B：怎么了？

A：We checked and found that there seemed to be a shortage. Are you sure you sent the full order?

A：我们检查后发现货少了。你确认全发货了吗？

B：How many are you short of?

B：少了多少？

A：We are short of five cases. The invoice shows forty-five units but we received only forty.

A：少了 5 箱。发票上是 45 箱。

B：That is terrible. We'll look into it immediately and send you the short part. As a compensation we'll deliver you one case more for free.

B：太糟糕了，我们马上补发，作为补偿多发一箱。这一箱免费送你。

A：Ok, thank you.

A：谢谢。

B：I hope the unhappiness will not affect our future cooperation.

B：我希望这场不愉快不会影响到我们未来的合作。

◇ Dialogue Ⅲ

这段对话发生在杭州红粉服饰有限公司外贸业务员 Dicky 与美国斯特朗服装有限公司采购部经理 Betty 之间，双方讨论货物在运输途中破损的问题。

A：美国斯特朗服装有限公司采购部经理 Betty

B：杭州红粉服饰有限公司外贸业务员 Dicky

A：Hi, Dicky.

A：Dicky 你好。

B：Was there a problem?

B：怎么了？

A：We had a damaged shipment. There was a lot of damage in transit.

A：我方的货物损坏了。在运输途中有很多损坏。

B：Could you describe the damage in detail?

B：能描述得详细点吗？

A：Two packing cases were crushed.

A：两个包装箱被压碎了。

B：Did you note the damage on the bill of lading?

B：你方在提单上在标注损坏了吗？

A：Yes, of course. I'd say about half of the shipment is unusable. Your company should pay us 50% as compensation fee.

A：当然。我想说大约一半的货物不可用，你方应该支付我方 50%的赔偿费。

B：The order was in good shape when it left our factory.

B：订单离开我们工厂的时候是好的。

A：What do you mean? It certainly didn't arrive here that way.

A：你什么意思？货到的时候不是好的。

B：I don't think this damage is our fault. It seemed the shipping company did this.

B：这不是我方的错。应该是航运公司的问题。

A：But the shipping company was recommended by you.

A：但是航运公司是你方推荐的。

B：I'm sorry. We'll look into it immediately and send you another packing cases with the price of 90% off. Do you agree with the deal?

B：很抱歉。我方会立即调查，再寄另一个九折的包装箱。你方同意这笔交易吗？

A：That's great. I'd contact the carrier to settle the damage.

A：太好了。我方会联系船运公司解决这批损失。

◇ Dialogue Ⅳ

这段对话发生在杭州红粉服饰有限公司外贸业务员 Dicky 与美国斯特朗服装有限公司采购部经理 Betty 之间，双方讨论货物品质与原合同不一致的问题。

A：美国斯特朗服装有限公司采购部经理 Betty

B：杭州红粉服饰有限公司外贸业务员 Dicky

A：Your last shipment is so disappointing that I cannot but file a claim against you.

A：你方上批货太让我方失望了，我方不得不向你方索赔。

B：What's wrong with it?

B：有什么问题吗？

A：Upon arrival of the goods at our port, we had them immediately reinspected. To our surprise, we found them far below the standard and they didn't meet the sample.

A：货物一抵达我方港口，我方立即重新检验。然而，令我们惊讶的是，我方发现它们远远低于标准，与样品不符。

B：Have you taken the picture of your sample? Will you let me have a look at it?

B：请问能拍个照片发给我看一下吗？

A：Of course!

A：当然了。

B：Have a look. Your dress is covered with lots of spots while there is no spot at all on our shipping sample. Compare these two samples, what conclusion can you draw yourself?

B：你看，你的衣服上全是斑点，而我方发运的货样却是洁白无瑕。比较这两件样品，你自己能得出什么结论呢？

A：We have noticed it already. These two samples do look different. But our cutting is taken from your arrived goods.

A：我方已经注意到这一点了。两个样品的确不一样。不过我方的样品正是从你方到岸的货物上剪下来的呀。

B：Our goods were strictly inspected by authorized department before loading. And the bill of lading was clean. This shows that the goods were in perfect condition and up to the export standard at the time of loading.

B：我方的货在装运前经过权威部门严格检验，提单也是清洁的。这说明货物装运时完好无损，符合出口标准。

A：Maybe the goods were damaged in transit.

A：货物有可能是在运输途中损坏的。

B：If you have evidence and your evidence is convincing, we shall meet the claim according to the international practice.

B：如果你方有令人信服的证据，我方将按国际惯例理赔。

A：All right. So be it at present.

A：好吧，暂时就这样吧。

◇ **Dialogue Ⅴ**

这段对话发生在杭州红粉服饰有限公司外贸业务员 Dicky 与美国斯特朗服装有限公司采购部经理 Betty 之间，双方讨论货物品质与数量与原合同不一致的问题。

A：美国斯特朗服装有限公司采购部经理 Betty

B：杭州红粉服饰有限公司外贸业务员 Dicky

A：Good morning, Dicky.

A：早上好，Dicky。

B：Good morning, Betty.

B：早上好，Betty。

A：Very regretfully, there occurred something unusual in this transaction of ours.

A：很遗憾，我们这次的买卖出了点意外。

B：What is the matter with you?

B：怎么了？

A：When your last shipment arrived at our port last week, we had them weighed. The result proved that the goods were underweight.

A：上一批货物上周抵达我方港口后，我方进行了过磅。结果表明货物短重。

B：Really?

B：真的吗？

A：As compared with the total weight of 100 metric tons stipulated in the contract, the actual landed weight was only 98.5 metric tons. The difference was 1.5 metric tons.

A：与合同规定的总重量 100 公吨相比，实际到货重量为 98.5 公吨。差异为 1.5 公吨。

B：That's really very incomprehensible. 1.5 metric tons is not small quantity and can't get lost on route. Where can these 1.5 tons have gone?

B：那真是没法理解了。1.5 公吨可不是个小数字，不可能在运输途中丢失呀。这 1.5 公吨到哪儿去了呢？

A：As the goods were sold on CIF terms, you must be held responsible for the shortage.

A：鉴于货物是按目的港交货价成交，你方要为短重负责。

B：Do you have any evidence?

B：你有证明吗？

A：Undoubtedly. Here's a survey report issued by a well-known notary lab in Singapore. It verifies the short weight of 1.5 metric tons. And then the report further testifies that the short weight is caused by improper packing. 15 bags were broken during transportation and the contents got irretrievably lost.

A：那当然了。这是新加坡一家著名的公证行签发的检验报告，证实短重 1.5 公吨。该报告还进一步证明短重是由于包装不当所造成。15 只袋子运输途中破损，袋中货物无可挽回地损失。

B：In view of our friendly relationship, we are prepared to meet your claim for the 1.5 metric tons short weight.

B：考虑到我们之间的友好关系，我方准备满足你方 1.5 公吨短重的索赔。

A：Do I understand that the inspection fee is also included?

A：是否检验费用也包括在内呢？

B：Yes.

B：是的。

Roles Simulation

Suppose you are one of them in the following conversation. Try to read aloud and practice the underlined sentences.

这段对话发生在杭州红粉服饰有限公司外贸业务员 Dicky 与美国斯特朗服装有限公司采购部经理 Betty 之间，双方讨论货物质量与原合同不一致的问题，商量具体赔偿事宜。

A：美国斯特朗服装有限公司采购部经理 Betty

B：杭州红粉服饰有限公司外贸业务员 Dicky

A：<u>There are two claims we make with you. Let's discuss them one by one.</u>

A：<u>我方向你们索赔有两点，我们一个个讲吧。</u>

B：Which one shall we discuss first?

B：那我们先谈哪一个呢？

A：The colors of last dresses severely fade. We have received the complaints from our customers. We cannot sell them out any more. <u>We have suffered a lot of loss.</u>

A：上次那批衣服色颜色掉得太厉害了。我方收到了顾客的投诉。这批货再也不能卖了。<u>我们的损失很大。</u>

B：Have you taken the picture of your sample? Will you let me have a look at it?

B：请问能拍个照片发给我看一下吗？

A：Of course! These are the pictures that our clients asked for the refund.

A：当然了。这些是顾客要求退款与退货的图片。

B：I'm so sorry to hear that. I hope you can take it easy. <u>I have to find out what caused the problem. If the problem was caused by us, we'll pay for your loss.</u>

B：很抱歉发生这种事。别着急，<u>我会调查具体的原因。如果是我方的问题，我方会赔偿你方的损失。</u>

A：Because of the bad quality, the dresses are unfit for sale again.

A：由于产品质量太差，这批衣服根本不适合销售。

B：This is most unfortunate. Our manufacturer has always attached great importance to the quality of their products.

B：真不幸发生这种事。工厂一直都狠抓质量的。

A：We have had long-good business relationship for almost ten years. <u>It was the first time that we found the goods far below the standard and they didn't meet the sample.</u>

A：我们是 10 多年的生意伙伴。<u>第一次碰到货品不达标，与样品不一致的情况。</u>

B：That's really very strange! <u>Whatever it is, we'll arrange the shipment to compensate your loss immediately when we check out the loss clearly.</u>

B：真奇怪。<u>不管怎样，等我方核实损失的具体情况后，立即安排重新发货。</u>

A：That's great!

A：太好了。

B：Can you tell me the number of the defective?

B：能告诉我次品的数量吗？

A：Ten pieces of dresses have been sent back to us. <u>The rest will be reinspected by our well-known notary lab in America, whose testimony is absolutely reliable.</u>

A：10 件已经退回我处了。<u>剩下的交由美国著名机构检验了。</u>

B：<u>It was rather a singular case. We have never come across such a case.</u>

B：<u>这是个特例，我方以前从来没有碰到过这种事。</u>

A：You'll get the survey report in two days.

A：两天后你方会拿到调查报告。

B：<u>That is an unfortunate oversight on our part and a lesson to us.</u> According to the survey report, we will pay the inspection fee.

B：<u>这个疏忽给了我们一个教训。</u>根据报告结果，我方会付检验费的。

A：<u>We accept your advice to settle the claim.</u> Well, in that case, there is nothing more to be said.

A：<u>我方接受这种处理方案。</u>这样的话，也没什么好说的了。

B：OK, Let's go into the next claim.

B：好吧，那我们谈下一个问题吧。

Typical Expressions

1. Upon the arrival of the last consignment in London, it was found, much to our regret, that about 50% of the cases were leaking. 最后一批货物到达伦敦后，令我们非常遗憾的是，大约 50%的箱子都有渗漏。

2. It is clear that you should take the responsibility for the excessive freight. 很明显，你方应该承担多余的运费。

3. As the quality of the goods was much lower than required, we therefore are lodging a claim of $1,000 for inferior quality. Your goods cannot be accepted as they differ from your samples. 由于货物的质量远低于要求，因此我们提出了 1,000 美元的劣质索赔。我方不能接受你方货物，因为这批货与样品不一样。

4. We will entertain your claim after the retained samples are rechecked and certified by the third authorized organization. 第三方官方授权机构重新鉴定留样结果出来后，我方会接受你方索赔的要求。

5. We will compensate your loss if we were responsible for it. 如果是我方责任，我方将赔偿你方全部损失。

6. It is the insurance company that you should ask for claim from, as the mishap happened after shipment. 你方应该向保险公司索赔，因为事故发生在装运后。

7. Your claim cannot be accepted because it is lodged 30 days after the arrival of the goods at the destination. 你方索赔不予受理，因为它是在货物到达目的地后 30 天提出的。

8. In view of the satisfactory conclusion of the matter, we will waive the claim for the inspection fee. 鉴于此事的结论令人满意，我们愿意撤回对检验费的索赔。

9. We greatly appreciate your cooperation in settling this unfortunate affair and look forward to a future extension of pleasant business relations. 我方非常感谢你方在解决这件事上全力配合，期待令人愉快的业务关系能进一步发展。

10. We regret that this unfortunate incident has ever occurred and hope it will not affect the business relations between us. 我们很遗憾发生这起事故，希望它不会影响我们双方的商业关系。

◇ Foreign Trade Documents

◇ Sample Inspection Booking

Intertek	INSPECTION BOOKING FOR ARMAND THIERY'S ORDERS	Date : 2021/05/12

Issue 1 - 140415

ARMAND THIERY
2, bis rue de Villiers
92309 LEVALLOIS PERRET

Dear Sirs,

We have received instructions from the above – mentioned principal for inspection of the following merchandise. **Please fill and return to us a copy duly signed with indication of**

1. **The date + time when goods will be available for inspection**
2. **Number of pieces per order which will be available for inspection**
3. **The address of the inspection location**

.

SUPPLIER (mentioned in L/C and quotation request) :
Name : xxxxxx.
Address RUE HENRI POINCARE ZONE D'ACTIVITE IRENE JOLIOT CURIE
City :   Country : FRANCE
Contact : xxxx
Tel : xxxxxxxxxxx
:

MANUFACTURER/ FACTORY (inspection place) :

Name : LIKFUNG KNITTING FACTORY(KC)
Address No.35-45,Ying Feng Rd. Min Ying Ind. Zone Dajingtou,
City :   Dong Guan   Country : China
Contact : xxxxxx
Tel : xxxxxxx
Inspection date + time requested : 2014/05/16 9:00 AM.

Armand Thiery 's purchase Order N°: KRM0556/AT

Quantities per order no : 5900 pcs

L.C. no:

Last shipment date : 2014/05/20

Product description : Ladies 60%viscose 40%polyamide knitted sweater

Documents Received   Reference Sample s available

Services Requested	Applicable Inspection Method
(√) Final Random Inspection (FRI)	ANSI / ASQC Z 1.4, sample size level II AQL critical 0 / Major 2.5 / Minor 4.0 defectives
( ) RE FRI	
Other services requested :_____	Special service instruction :_____

Supplier Signature :   Martin Mao

Oral Practice

Forming a group of two or three partners. Try to work out a dialogue related to the theme of this unit and then perform it in class.

Self-Assessment

Evaluate your practice by marking the following each corresponding item and work out the total scores.

考核情境(分值)	考核要求(分值)	得分
语音语调(20 分)	发音准确，声音清晰(10 分)	
	语调自然，语速自然流畅(10 分)	
内容(30 分)	内容完整(10 分)	
	内容表述符合给定情境(10 分)	
	前后内容逻辑一致，内容之间无矛盾(10 分)	
语言描述(30 分)	用词规范、准确，无语法错误(20 分)	
	表述无歧义(10 分)	
其他(20 分)	无雷同状况(10 分)	
	无严重偏题(10 分)	
合　　计(100 分)		

Module II

Cross-Border Electronic Commerce Trade

Task 1   E-store Introduction

知识目标

◆ 学习跨境电子商务中平台店铺介绍的内容
◆ 学习跨境电子商务中平台店铺介绍的艺术
◆ 学习相关商务沟通的书面英文表达方式

能力目标

◆ 能够准备平台店铺介绍材料
◆ 能够顺利开展平台店铺介绍活动
◆ 能够处理平台店铺的往来信函

Basic Knowledge Review

1. What is e-commerce and what are basic types of e-commerce platforms?

E-commerce is a modern-day invention that facilitates the trading of goods and/or services through electronic means, or more precisely, the internet.

The following are the different types of e-commerce platforms:

(1) Business-to-Business (B2B)

A B2B model of business involves the conduct of trade between two or more businesses/ companies. The channels of such trade generally include conventional wholesalers and producers who are dealing with retailers.

(2) Business-to-Consumer (B2C)

Business-to-Consumer model of business deals with the retail aspects of e-commerce, i.e. the sale of goods and/or services to the end consumer through digital means. The facility, which has taken the business world by storm, enables the consumer to have a detailed look at their proposed procurements before placing an order. After the placement of such orders, the company/agent receiving the order will then deliver the same to the consumer in a convenient time-span. Some of the businesses operating in this channel include well-known players like Amazon, Flipkart, etc.

This mode of purchase has proved to be beneficial to the consumers when compared to the traditional method, as they are endowed with access to helpful contents which may guide their purchases appropriately.

(3) Consumer-to-Consumer (C2C)

This business model is leveraged by a consumer for selling used goods and/or services to other consumers through the digital medium. The transactions here are pursued through a platform provided by a third party, the likes of which include OLX, Quickr, etc.

(4) Consumer-to-Business (C2B)

A C2B model is the exact reversal of a B2C model. While the latter is serviced to the consumer by a business, the C2B model provides the end consumers with an opportunity to sell their products/services to companies. The method is popular in crowdsourcing based projects, the nature of which typically includes logo designing, sale of royalty-free photographs/media/design elements, and so on and so forth.

(5) Business-to-Administration (B2A)

This model enables online dealings between companies and public administration, i.e. the Government by enabling the exchange of information through central websites. It provides businesses with a platform to bid on government opportunities such as auctions, tenders, application submission, etc. The scope of this model is now enhanced, thanks to the investments made towards e-government.

(6) Consumer-to-Administration (C2A)

The C2A platform is meant for consumers, who may use it for requesting information or posting feedbacks concerning public sectors directly to the government authorities/administration. Its areas of applicability include:
- ✓ The dissemination of information.
- ✓ Distance learning.
- ✓ Remittance of statutory payments.
- ✓ Filing of tax returns.
- ✓ Seeking appointments, information about illnesses, payment of health services, etc.

2. What is an e-store?

An e-store may also be called an e-web-store, e-shop, an online store, Internet shop, web-shop, web-store, online storefront and virtual store. Mobile commerce (or m-commerce) describes purchasing from an online retailer's mobile device-optimized website or software application ("app"). These websites or apps are designed to enable customers to browse through a companies' products and services on tablet computers and smartphones.

3. What is customer buying behavior in digital environment?

In the marketing around the digital environment, customer's buying behavior may not be influenced and controlled by the brand and firm, when they make a buying decision that might concern the interactions with search engine, recommendations, online reviews and other information. With the quickly separate of the digital device environment, people are more likely to use their mobile phones, computers, tablets and other digital devices to gather information. In other words, the digital environment has a growing effect on consumer's mind and buying behavior. In an online shopping environment, interactive decision may have an influence on aid customer decision making. Each customer is becoming more interactive, and though online customers reviews can influence other potential buyers' behaviors. In addition, not only those reviews, people more rely on other people's post information about product commends on social media. Common problems in the past and some solutions or comments of the merchants will be attached for customer reference.

People cannot examine whether the product can satisfy their needs and wants before they receive it. Customer may concern after-sale services. Finally, customer may afraid that they cannot fully understand the language used in e-sales. So trust is another way driving customer's behaviors in digital environment, which can depend on customers' attitude and expectation. Indeed, the company's products design or ideas cannot meet customer's expectations. Customer's purchase intention is based on rational expectations and emotional trust.

4. How many e-commerce fulfilment models?

There are three main fulfillment models associated with e-commerce that dictate the role of the retailer as well as the way in which a product is stored and distributed to the end-user. These models have a significant impact on the operational characteristics of the business and its day-to-day running as well as the overall operating margin. The three main models are:

(1) Dropshipping Model

In a dropshipping model, the e-commerce business takes no physical possession of the items on sale. The store owner does not keep products in stock and there is no inventory held. Instead, orders are sent directly to the manufacturer, who is responsible for storing the items and shipping them to the customer. In this sense, the merchant never sees or touches the products, which has some unique advantages over adopting a more traditional order fulfilment model.

(2) Traditional Order Fulfilment Model

Buying wholesale is arguably closest to the traditional offline retail model. In effect, the business owner (retailer) acquires stock directly from a wholesaler at a discounted rate, applies a margin onto each product and decides to deliver to consumers directly.

(3) Outsourced Fulfilment Model

The retailer may wish to market products from a supplier that does not provide a dropshipping service. If the retailer wants to avoid end-to-end fulfilment (like the traditional model), then a hybrid approach can be adopted—using a "fulfilment house". In this model, companies such as Shipwire are commissioned to handle the product side of the business, on behalf of the retailer. Generally speaking, they are responsible for collecting products from the supplier, holding the product at their distribution centers, all packaging as well as onward order fulfilment (to the customer). This service comes at a cost—minimum fees, return fees and setup fees are commonplace and should be fully weighed up beforehand.

Task Descriptions

杭州红粉服饰有限公司跨境电子商务部运营专员 Lily 开始筹备跨境电子商务平台店铺关于公司概况、主营业务介绍等事宜，并查阅参考资料。

E-store Introduction

Typical Samples

◇ **Sample Ⅰ   Company profile**

Business Type: Manufacturer/Factory & Trading Company

Business Range: Apparel & Accessories, Textile

Main Products: Sweater, Scarf, Cashmere Sweater, Knitwear, Cardigan, Knitted Pullover, Cashmere, Wool

Year of Establishment: 1998-01-12

Terms of Payment: LC, T/T, D/P

Management System Certification: ISO 9001, ISO 9000, BSCI, WRAP

Main Markets: North America, Eastern Europe, Mid East, Eastern Asia, Western Europe

◇ **Sample Ⅱ   Production capacity**

Factory Address: Zhongshan Road, Haitian District, Hangzhou, Zhejiang, China

R&D Capacity: OEM, ODM, Own Brand

No. of R&D Staff: 5-10 People

No. of Production Lines: 4

Annual Output Value: US$50 Million - US$100 Million

■ **Notes:**

1. ODM: ministry of overseas development 海外开发部

2. OEM: original equipment manufacturer 原始设备制造商

◇ **Sample III   Trade capacity**

International Commercial Terms (Incoterms): FOB, CIF

Terms of Payment: LC, T/T, PayPal

Average Lead Time: Peak season lead time: 1-3 months / Off season lead time: 1-3 months

Number of Foreign Trading Staff: >50 People

Export Year: 2003-09-13

Export Percentage: 80%-95%

Main Markets: North America, Eastern Europe, Mid East, Eastern Asia, Western Europe

Nearest Port: Shanghai, Ningbo

■ **Notes:**

lead time: 订货至交货的时间

◇ **Sample IV   Company introduction   I**

Dongguan Knitted Garment Co., Ltd was established in March 2003. Our company is located in Guangdong Province. Our company started with a small-size factory, running in rented plant. In 2004, we were engaged in international trade and established our own imported & exported company.

We are one of the professional knitted garment manufacturers. We have been focusing on making wool products for over 18 years. We provide service from yarn ordering, yarn knitting, fabric finishing, sewing, ironing & packing as a whole. We are experienced in stitch typing, pattern making and garment designing.

Our company' goal is to keep hunger, stay focus. We have always been keeping eyes on quality control. We have been striving to make the communication with clients more efficient and transparent. With the rapid development, we have attracted foreign customers around the world, and gained high reputation.

Great customer service is provided by our experienced merchandisers with fluent English & rich fashion knowledge. Instant feedback and effective communication can ensure the order be finished correctly & effectively.

◇ **Sample Ⅴ Company introduction Ⅱ**

Jiaxing knitting group Co., Ltd was established in 1992. Our company is an international trade enterprise. We specialize in manufacturing and exporting various sweaters made of silk, cotton, wool, cashmere and blended materials. We sell knitting fabrics and textiles as well.

For more than twenty years' development, our company has become one of the top knitting enterprises in Hangzhou. In 2008 our sales revenue reached 500 million RMB. Now we can make 71 million knitting sweaters each year.

Jiaxing knitting group Co., Ltd is always committed to making the perfect products for customers. To obtain this goal, we lay much attention to every procedure in clothing making. We have well-trained professionals in charge of quality-inspection before all the products are ready to be delivered.

We have been improving production capacity and technology. In 2010 we made a structural adjustment in order to make task-distribution clearer and our production lines work more smoothly. So far, we have 400 sets of advanced semi-auto manual knitting machines, 50 sets of auto computer machines and 5,678 sets of manual knitting machines suitable for various gauge sweater making.

As a large and famous enterprise, because of our good reputation on quality and service, we have won many customers worldwide especially in our main market, USA and EU.

✿✿ *Typical Expressions*

1. Our company was established in 2010.We are located in Hangzhou and covers an area of 2,000 square meters. We have around 200 staff members. 公司成立于 2010 年，总部在杭州，占地面积 2,000 平方米，员工大约 200 人。
2. Our products include cashmere-wool knitted and woven blanket, cushion cover, quilt, slippers, socks, robe and travel set, etc. 我们的产品包括羊绒毯、垫套、被子、拖鞋、袜子、长袍、旅行套装等。
3. Jeans are our main project. / We specialize in all various jeans. 牛仔裤是我们的主营项目。
4. We have 14-year business experience in bags, shoes and their accessories in China, and we run our own factory with more than 100 employees. 我们在中国经营鞋包及配饰已有 14 年了，

拥有自己的工厂，员工人数超过 100 人。

5. In order to provide our customers with top quality and innovative products we use stringent quality control measures.为了给顾客提供质量上乘的产品，我们采用了严格的质量管理措施。

6. We have employed experienced technical and management personnel. 我们拥有一大批优秀的技术人员和管理人员。

7. Our store has been awarded 3 diamonds, which means we have received more than 2,000 positive feedbacks from our customers around the world. 店铺已经上了三颗钻石，这表明我们的全球顾客好评有 2,000 多个了。

8. Quality is superior. Service is supreme. Reputation is first. 质量上乘，服务优质，声誉第一。

9. Our aim is to provide you products with high quality and competitive price. 为您提供价廉物美的产品是我们奋斗的目标。

10. Mutual development, mutual benefits. 共同发展，共创双赢。

Composition-Translation

◇ **Part Ⅰ   Integrating the following Chinese basic information into English.**

杭州百合服饰跨境电商有限公司

(Hangzhou Lily Fashion Cross-Border E-commerce Limited Company)

杭州百合服饰跨境电商有限公司简称百合服饰(Lily Fashion)成立于 2008 年 9 月，总部在中国杭州。百合服饰是一家专营女装、童装的店铺。百合服饰主打产品为针织衫、婴儿装、孕妇装、职业裙装和四季连衣裙等，服饰做工精细，面料亲肤、柔软和透气。公司拥有自己的研发设计团队，自主品牌，客源地主要为北美洲和欧洲。

◇ **Part Ⅱ   Integrating the following English basic information into Chinese.**

Hubei Grand Cashmere Co. Ltd

Hubei Grand Cashmere Co. Ltd is a manufacturer which produces cashmere, wool scarf shawls and other accessories. We are also an exporter which export cashmere, sheep cashmere, wool woven and knitted scarf shawls, gloves, hat sets and worsted pashmina shawls which are solid, reversible, checked, graded, printed, jacquard and knitted. Our head company is a raw-dehairing cashmere factory, which dehairs goats' wool into pure cashmere fibers to produce scarf, shawls and knitwear etc. The cashmere wool raw materials are all from inner Mongolian grassland.

Clients can choose from our ready styles, or send ideas of your own design to develop. Clients can also send us the style samples and pictures with details about sizes and dimensional sketch. Clients usually should make sample confirms with us before bulk order.

Welcome you visit our company, and hope to cooperate with you.

Self-Assessment

Evaluate your practice by marking the following corresponding each item and work out the total scores.

考核情境(分值)	考核要求(分值)	得分
行文逻辑(20分)	层次清楚，思路清晰(10分)	
	行文流畅，通顺(10分)	
内容(30分)	内容完整，涵盖全部写作要点(15分)	
	内容表述精准，忠于原文(15分)	
语言描述(30分)	用词规范、准确，无语法错误(20分)	
	句式富于变化(10分)	
其他(20分)	无雷同状况(10分)	
	无严重偏题(10分)	
合　计(100分)		

Task 2   Product Description

知识目标

◆ 学习跨境电子商务中产品描述的内容
◆ 学习跨境电子商务中产品描述的艺术
◆ 学习相关商务沟通的书面英文表达方式

能力目标

◆ 能够准备产品描述材料
◆ 能够顺利开展产品描述活动
◆ 能够处理产品描述的往来信函

Basic Knowledge Review

1. What are product descriptions?

As the name suggests, product descriptions are the words you use to describe your products. In ecommerce, specifically, they're the blurbs of text on product pages that tell customers about your product. A good product description describes your product's features and benefits, acknowledges the problem it solves, and declares why it's the best product for the job.

Strategic, branded product descriptions can be the difference between an abandoned cart and a conversion. While you might save time by pasting generic descriptions into your product pages, taking the path of least resistance can really limit your bottom line.

It's true that product descriptions take time to write—and even more time to write well—but they're a critical part of merchandising your products. Being able to write clear, compelling product descriptions is a huge part of running a successful online store.

2. Do customers need product descriptions?

Yes, they do. Your customers need them.

A product listed without a description will sell at a much lesser rate than a product with one. Imagine trying to sell a waterproof jacket without saying it's waterproof, or trying to sell a pair of eco-friendly shoes without saying they're made of recycled materials. Good luck.

Brick-and-mortar shops have employees available to provide customers with basic product information. Plus, as a shopper, you can physically interact with a product and get more information before you decide to buy it.

Online, it's very different. In online stores, product descriptions try to replace the experience of browsing and interacting with in-store products. When done well, they take away the guesswork and ensure customers aren't left wondering what a product is really like.

3. How does the e-shop owner write product descriptions?

Here are some steps to help the e-shop owner to write product descriptions.

(1) Figure out your target demographic

Start by asking yourself: who is going to buy this product? Once you have an idea of who your target customer is, you can craft your product description to fit that person's lifestyle, interests, and needs. If you're selling baby toys, you'll want your descriptions to appeal to new parents rather than childless university students. Determining customer profiles can also help with marketing, product development, and store design.

(2) Gather basic information

What are the basic pieces of information you want to convey in your product description? If you're selling clothes, you may want to include any pertinent information on materials, manufacturing process, place of origin, etc. Does a garment fit small? Does it use natural dyes? Product descriptions should cover any basic information that can't be answered by photos.

(3) Keep your language accessible and cohesive

While your product descriptions may be intended for a specific audience, you still want to ensure anyone who visits your store can understand the product. Everything should flow in a logical order. If it doesn't make sense to you, the expert, your customers definitely won't know what you're talking about.

(4) Don't forget SEO

SEO stands for Search Engine Optimization, and it's something you'll want to keep in mind. Strategically written product descriptions include keywords that make it easier for search engines to find your products.

(5) Consider FAQs

What are your customers' most frequently asked questions?

✓   Can this go in the dishwasher?

✓   What will happen if I forget this in the trunk of my car for a month?

✓   Is this a good present for my mom?

✓   Was this sustainably made?

Making a list of your frequently asked questions and including answers in your descriptions will make life easier for you and your customers. With proper product descriptions, you can expect customer support inquiries to decrease. Get in the habit of combing through your FAQs and see if there's any information that you can add to your product descriptions as time goes on.

Task Descriptions

杭州红粉服饰有限公司跨境电子商务部运营专员 Lily 对上架产品信息进行编辑完善。

Product Description

Typical Samples

◇ Sample Ⅰ　Dress

Product Name:

Women's V-Neck Floral Spaghetti Strap Summer Casual Swing Dress

Search Terms:

Women's Summer Beach long sleeveless sex string sunflower black bottom dress

Product Description:

This summer dress wear is a flawless addition to any collection. This sleeveless floral dress comes with V neck two side pockets, and elastic waist. This is an essential part of your wardrobe.

Type: Floral Sleeveless Dress, Casual Dress, Summer Beach Dress

Pattern Type: Floral printed

Neckline: V-Neck Sleeve

Length: Sleeveless/Spaghetti Strap

Season: Perfect for Spring, Summer and Autumn

Occasion: You could wear it with high heels, flats, sandals or pumps. Perfect for beach or outside.

Newstyler　Pinklife Size Chart (in)				
Size	Length	Chest	Waist	Bottom
Small	40.6	27.2	26.0	59.1
Medium	40.9	28.7	27.6	60.6
Large	41.3	30.3	29.1	62.2
X-Large	41.3	31.9	30.7	63.8
P.S. Please choose right size according to your dressing habits such as slim or loose fit. If you like loose fit, please choose a larger size than the reference.				

◇ Sample II Skirt

Product Name:

Women Stylish Floor Length Floral Printed Maxi Circle Skirt

Product Tag:

Women Dress/Skirt/ Maxi Skirt/ Floor Length Skirt/ Stylish Skirt

Product Description:

Waist type: high and circle waist

Main fabric component: polyester and spandex

Size: S, M, L, XL, XXXL

Size	Bust	Waist	Length	US
S	86-110CM/33.86-42.9"	74-96CM/29.13-37.44"	140CM/55.12"	4-6
M	90-114CM/35.43-44.46"	78-100CM/30.71-39"	141CM/55.51"	8-10
L	94-118CM/37.01-46.02"	82-104CM/28.28-40.56"	142CM/55.91"	12-14
XL	98-122CM/38.58-47.58"	86-108CM/33.86-42.12"	143CM/56.3"	16-20

Attention: Please allow 1-3cm error because measured by hand. Please refer to the sizing chart before choosing. Due to the difference between different monitors, the picture may not reflect the actual color of the item.

This is a lovely long length maxi dress. Made of 90% polyester and 10% spandex. It has soft material and has a stretch to it.

It has an elastic waist, floor length and available in four color patterns. Also great for maternity wear!

◇ Sample III Socks

Product Name：Butterfly Design Cotton Crew Socks

Product Tags: Socks/ Men's Socks /Women's Socks /Short Socks/ Spring Socks/Cotton Socks/Winter Socks/Fall Socks/Summer Socks

Product Description:

Style: Casual

Material: Cotton

Season: All Seasons

Fabric: High Stretch

Fit Type: Regular Fit

Size: One size

How to measure: This data was obtained from manually measuring the product, it may be off by half inch or 1-2 CM.

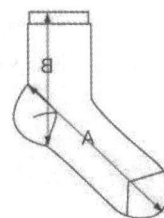

Size Chart				
	A：FOOT LENGTH		B：SOCK LENGTH	
SIZE	CM	INCH	CM	INCH
one size	21	8 1/4	21	8 1/4

◇ Sample Ⅳ Shoes

Product Name：Men Sandals Summer Casual Shoes Soft Mesh Couple Beach Slippers
Foam-Runners

Product Tags: Men Sandals/ Casual Shoes/ Beach Slippers/
Runners/ Mesh Slippers

Product Description:

Style: Ethnic Fashion

Occasion: Casual

Upper Material: Flexible Rubber Plastic

Color: Black/White/Red/Beige

Size:

Inner Length(mm)	143	156	169	182	195	208
Size	22/23	24/25	26/27	28/29	30/31	32/33
Inner Length(mm)	221	234	247	260	273	286
Size	34/35	36/37	38/39	40/41	42/43	44/45

Oak upper touches delicate and smooth.

Soft bending without crease. Fashion and attractive leading design.

Oak flat durable soles with concave and convex texture are fit for all weathers such as heavy rain.

◇ Sample Ⅴ Wallet

Product Name: Wallet Luxury Designer Large Capacity Coin
Card Phone Clutch Holder Bags Pouch Handbag Purse

Product Tags: Wallet/ Bags/ Pouch Handbag/ Coin Purse/ Phone
Bag

Product Description:

Item Type: Classic Wallet

Main Material: PU

Size: 10 × 8 × 3cm

Weight: 80g

Gender: Unisex

Closure Type: zipper

Decoration: Sequined

Interior: Interior Slot Pocket/ Cell Phone Pocket/Interior Zipper Pocket/Coin Pocket/Card Holder

Typical Expressions

1. product name 产品名称/标题　　　　　　product tags 标签
   product description 产品描述　　　　　　color 颜色
   volume 容积　　　　　　　　　　　　　specifications 规格
   length 长度　　　　　　　　　　　　　width 宽度
   size 尺码　　　　　　　　　　　　　　weight 重量
   category 类目

2. color 颜色：

black 黑色	green 绿色	ivory 象牙白
gray 灰色	pink 粉色	navy blue 海军蓝
red 红色	brown 棕色	orange 橙色
purple 紫色	blue 蓝色	tan 黄褐色
yellow 黄色	gold 金色	white 白色
multi-color 混色	beige 米黄色、淡棕色	bronze 深红棕色

3. Wish 平台类目：

Hobbies 个人兴趣爱好产品	Gadgets 电子产品
Shoes 鞋子	Fashion 衣服
Automotive 汽车内饰配件	Tops 上衣
Bottoms 休闲运动裤	Underwear 内裤
Watches 手表	Wallets & Bags 包包
Accessories 小饰品	Phone Upgrades 手机配件
Home Decor 家居装饰	Sports & Outdoors 户外运动产品

4. Material: Straps and panel: 100% Organic Cotton　材料：肩带和面板：100%有机棉

5. Washable: Up to 30℃, Gentle Cycle, Slow Spin Speed 可水洗：最高温度 30℃，轻柔

6. Color: Dark Blue, Purple, Turquoise, Pink 颜色：深蓝色，紫色，绿松石色，粉红色

7. Fully adjustable and suitable from 3 kg up to 18 kg/7 - 40 lbs. 带子可调节，适用于 3 公斤至 18 公斤/7-40 磅

8. Made in China and neighboring countries. 产地中国及周边国家。

9. It allows you to carry your baby on your front, hip and back at the appropriate ages while your child is always supported across the back and the bottom ensuring the natural spread squat position. 这款婴儿座可以根据孩子的年龄任意调节带子，把孩子放在大人的前面、臀部和

背部，并且保证孩子在背部和臀部获得支撑并保持自然展开蹲的姿势。

10. Aircraft alloy shank, light strong, corrosion proof. Solid handle for maximum comfort and torque. 飞机合金刀柄，轻强度，耐腐蚀。坚固的手柄，提供最大的舒适性和扭矩。

Composition-Translation

◇ Part Ⅰ   Integrating the following Chinese basic information into English.

产品名称：儿童女孩泳衣两件套泳衣泳裤

材料：聚酯

型号：S324

适合性别：女孩

场合：游泳，沙滩

特点：高弹性，可爱，时尚

风格：儿童防护

产地: 中国

尺码：4A/6A/8A/10A/12A/14A

颜色：粉色

推荐年龄	尺码	胸围	衣长	袖长	裤子腰围	裤长
3~4 岁	4A	58cm/22.8"	40cm/17.7"	6.75cm/2.66"	48cm/18.9"	21cm/8.3"
5~6 岁	6A	61cm/24.0"	42cm/16.5"	7.00cm/2.76 "	50cm/19.7"	23cm/9.1"
7~8 岁	8A	64cm/25.2"	45cm/17.7"	7.25cm/2. 85"	52cm/20.5"	24cm/9.4"
8~9 岁	10A	68cm/26.8"	49cm/19.3"	7.50cm/2.95"	54cm/21.3"	25cm/9.8"
9~10 岁	12A	72cm/28.3"	52cm/20.5"	7.75cm/3.05"	56cm/22.0"	26cm/10.2"
11~12 岁	14 A	75cm/29.5"	55cm/21.7"	8.00cm/3.15"	59cm/ 23. 2"	28cm/ 11. 0"

◇ Part Ⅱ   Integrating the following English basic information into Chinese.

Product name: Autumn Dress French A- line Skirt Vintage Floral Puff Sleeve Lace-up Waist-Controlled Slimming Sweet Pleated Midi Dress

Silhouette: Straight

Season: Spring/Autumn

Neckline: O-Neck

Material: COTTON

Age: Ages 18-35 Years Old

Sleeve Length(cm): Full

Material: Polyester

Dresses Length: Knee-Length

Closure Type: Lace

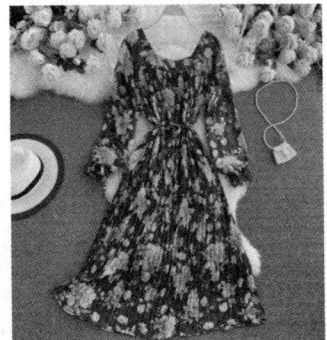

1. All measurements are listed in cm (1cm = 0.39 inch). There is a 2-3cm difference according to manual measurement. Please check the measurement chart carefully before you buy the item.

2. Due to the light and screen, the slight color difference should be acceptable.

Thank you for your understanding !!!

Suggestion: Choose the size according to your weight.

Size S - Weight: 40 kg - 45 kg

Size M - Weight: 47.5 kg - 52.5 kg

Size L - Weight: 55 kg - 60 kg

Size XL - Weight: 60 kg - 62.5 kg

Free size - Weight: 45kg - 60 kg

Self-Assessment

Evaluate your practice by marking the following each corresponding item and work out the total scores.

考核情境(分值)	考核要求(分值)	得分
行文逻辑(20 分)	层次清楚，思路清晰(10 分)	
	行文流畅，通顺(10 分)	
内容(30 分)	内容完整，涵盖全部写作要点(15 分)	
	内容表述精准，忠于原文(15 分)	
语言描述(30 分)	用词规范、准确，无语法错误(20 分)	
	句式富于变化(10 分)	
其他(20 分)	无雷同状况(10 分)	
	无严重偏题(10 分)	
合　计(100 分)		

Task 3   On-Sale Customer Service

知识目标

◆ 学习跨境电子商务中售中客户服务的内容
◆ 学习跨境电子商务中售中客户服务的艺术
◆ 学习相关商务沟通的书面英文表达方式

能力目标

◆ 能够准备售中客户服务材料
◆ 能够顺利开展售中客户服务活动
◆ 能够处理售中客户服务的往来信函

Basic Knowledge Review

1. What is e-commerce on-sale customer service?

E-commerce customer service is how online businesses provide assistance to customers with everything from making online purchase decisions to resolving issues — all while creating a seamless customer experience across channels and platforms.

In this digital-first world, e-commerce customer service is not simply nice to have, but rather a prerequisite for success. Data from Microsoft shows that for 95% of consumers, customer service is important for brand loyalty.

However, it's not enough to say that you have customer service. Bad customer service is worse than none at all. There is a significant discrepancy in the perceptions of companies and buyers when it comes to the quality of service, as 80% of businesses believe they provide excellent customer service, but only 8% of customers agree.

Today's customers have sky-high expectations. Even though fewer customers may be experiencing problems, more customers are inclined to complain about customer service problems than ever before.

The good news is that Millennials are ready to pay 21% more to do business with companies who excel at customer service.

2. How to improve the on-sale service as an effective e-commerce operator?

The biggest drawback of online shopping is the time gap between purchasing and playing with goods. Here are several ways you can step up your ecommerce customer service.

As customers' expectations for the ideal online shopping experience increase, customer service becomes essential to thrive in a competitive e-commerce landscape.

(1) Get organized

Keeping track of customer's conversations and equipping your team with tools that help them collaborate keeps everyone on the same page.

(2) Meet customers on their terms

In e-commerce, one size fits none. Customers expect a personalized approach that makes them feel special. Businesses today have to interact with their consumers to get to know them better and form meaningful, ongoing relationships.

(3) Stand out from the crowd using personalization

Unlike situations when customers just want quick answers to basic questions, there are times when they are looking for a personalized approach and expert advice. Introducing a personalized, knowledgeable service is what sets leaders apart from other online businesses.

(4) Harness the power of customer reviews

Today's customers are more empowered than ever to make informed decisions. They want to be heard, and they want to hear other customers' opinions about products and services. Besides the fact that reviews are among major purchase decision factors, they're also a great source of ideas for customer experience improvement.

(5) Improve your response time

E-commerce is all about speed and convenience. Customers who choose to shop online are expecting a fast reaction and prompt answers. Allowing customers to reach your customer service team using different channels is not enough. The key to a well-performing team is the balance between speed and convenience.

When you're able to serve customers with a consistent level of quality in a short period of time across different channels, you're on the path to providing a fully functional, multichannel customer service strategy.

(6) Measure, optimize, repeat

No customer service approach is set in stone. Companies have to be proactive in solving customers' problems while constantly improving processes based on data. Evaluating your team's volume by channel, tracking busiest hours, and following trending topics among your customers are just some of the things that you can refine by measuring and optimizing results. Having a process in place to track performance will serve as a foundation for making future decisions, which is the first step toward successful customer service.

Task Descriptions

杭州红粉服饰有限公司跨境电子商务部运营专员 Lily 在线处理跨境电商平台客户下单前关于产品、物流、价格和支付等售中事宜。

On-sale Customer Service

Typical Samples

◇ Sample Ⅰ

由于技术层面的原因，Amazon Customer Service 后台自动将订单做取消处理，Lily 不得不执行退款操作。下面是 Amazon Customer Service 发给买家关于取消订单的原文。

> Hello Sandy，
> we regret to inform you that your order to Congo was processed at a discounted price due to a technical issue. As a result, the other will be automatically cancelled. Such occurrences are rare and we are sorry for the inconvenience. If you have paid for your order, the amount will be refunded to the original payment method used once the order is cancelled. Refunds can take up to 3～5 business days to reflect on your account depending on your bank or card issuer's policy.

◇ Sample Ⅱ

买家付款后，作为卖家的 Lily 给客户发了封通知信。下面是通知信函的原文。

> Dear Gee,
> We would like to inform you that your payment has been received and Order No. NJ0019 being processed.
>
> Once the items have been packed, they will be shipped to you immediately. And you can view the status at any time.
>
> If you have any question on your payment, please feel free to contact us.
>
> Hope you enjoy shopping with us!
> Sincerely,
> Pinklife Team

◇ **Sample Ⅲ**

针对买家还未收到货品，Lily 立即与物流商核实物流信息，并将调查结果及时告知客户。下面是回函的具体内容。

Hello,

I'm sorry for the inconvenience you've had with this purchase. I understand that you still haven't received your order. I understand how inconvenient this is for you. I'll be pleased to help you with the problem.

I tried to track your package. After contacting the carrier, I've confirmed it was shipped out and it is now in transit to be delivered to you. I'm sorry about the delay. However, please understand that some delays can be caused by external factors. No matter how careful we try to be, unexpected events can still happen which are beyond our control.

Would you please wait for 3 more days? If you still don't get the delivery, please contact us. Thank you for your understanding.

Yours,
Pinklife Team

◇ **Sample Ⅳ**

针对买家今天刚刚下的订单，Lily 立即给买家写了封感谢信。下面是感谢信的具体内容。

Dear John,

Thank you for choosing us for your purchase. We are happy to meet your purchase order and value our loyal customers.

You have ordered the dress from us today. As a way to show our gratitude we offer you a 20% off coupon for your next purchase.

Regards，
Pinklife Team

◇ **Sample Ⅴ**

海外买家通过阿里国际站找到杭州红粉服饰有限公司联系方式，打电话来询问价格。下面是 Lily 与客户的对话内容。

A：海外买家

B：杭州红粉服饰有限公司客服 Lily

A：I'm interested in this cashmere coat. The price of $300 is on the high side. Could you charge a bit less?

A：我对这件羊绒外套很感兴趣。300 美元的价格很高。你能少收一点费用吗？

B：Sorry. This is the latest style this spring. It's the best-seller at my store.

B：对不起。这是今年春天的最新款式。这是我店里的畅销书。

A：Mother's Day is coming in two weeks. Do you have any promotions for it?

A：母亲节在两周后就要到了。你有什么促销活动吗？

B：Yes. Our promotion begins next week to celebrate Mother's Day. We'll offer discounts up to 50% off, $5, $10 and $20 coupons, and some special priced products. We'll also add some sweet Mother's Day gifts for every order.

B：是的。为庆祝母亲节，我们的促销活动将于下周开始。我们将提供 5 折的折扣，5 美元，10 美元和 20 美元的优惠券，以及一些特价的产品。我们还会为每个订单赠送甜蜜的母亲节礼物。

A：Wow! That sounds fantastic! Will this coat be on sale?

A：哇！这听起来太棒了！这件外套到时在售吗？

B：Yes. You can get 30% off for it, and a $10 coupon for your next purchase within one month.

B：是的。你可以享受 30%的折扣，并获得有效期为一个月的 10 美元优惠券。

A：I really look forward to that. What's the gift?

A：我真的很期待着这个活动。礼物是什么？

B：It's a secret now. But I'm sure that you'd love it.

B：现在这是个秘密。但我相信你会喜欢它的。

💞 Typical Expressions

1. Are you online/there/here? 您在线吗？

2. Welcome to my shop. May I help you? /Can I help you?/What can I do for you? What are you interested in? 欢迎光临，需要帮忙吗？您想买什么？

3. Please leave me a message if you have any questions. I'll reply to you as soon as I'm back. 如果您有问题请留言，我回来后会尽快给您回复。

4. Sorry, the dress you wanted is not in stock/out of stock. 很抱歉，您要的这款裙子没有库存了。

5. Can you tell me what the leggings are made of, artificial leather or genuine leather? 请问这款打底裤用的是什么材料，人造皮革还是真皮？

6. I wonder if the white dress is in stock or out of stock. I am 175cm, which size fits me? 这款裙子有白色的吗？身高 175 厘米选哪个码合适？

7. The rain pants are very waterproof but not vey thick and they will keep your legs dry. 这款雨裤不但防水性能好，而且轻薄。可以防止腿部被雨水打湿。

8. The manufacturer stopped producing the shapewear last month. 这款塑身衣厂家上个月就停止生产了。

9. Can you recommend some items to me? 您能给我推荐一些款式吗？

10. These styles I just sent you are winter best-sellers. 我刚发给您的这些款式都是冬季的畅销款。

Composition-Translation

◇ Part Ⅰ Integrating the following Chinese basic information into English.

库存不足的反促销信

亲爱的约翰，

感谢您再次光临本店。很抱歉，您上次购买的打底裤已经断货了。您要不要看看其它面料款式的打底裤？

本月 10～13 号是我们周年店庆日，为此店里新推出一批款式新价格合理的裙装，适合各种场合。

光顾本店，享受全店八五折优惠。另外，如果一次性消费 600 美元以上，您就可以享受二折的优惠，如果您一次性消费 800 美元以上，您可以享受三折优惠。

本次促销活动 13 日结束，希望您购物愉快！

您的，
Pinklife Team

◇ Part Ⅱ Integrating the following English basic information into Chinese.

The buyer hasn't made payment

Dear Madam,

We noticed that you haven't paid for the order. This special deal will be ending in 3 days. If you place an order before the deadline, we will send some nice gifts with your order. You know, this is the biggest year-round promotion this year, with the price of some items even up to 90% off. If you need any help or have any questions, please let us know.

Best regards,
Pinklife Team

Self-Assessment

Evaluate your practice by marking the following each corresponding item and work out the total scores.

考核情境(分值)	考核要求(分值)	得分
行文逻辑(20 分)	层次清楚，思路清晰(10 分)	
	行文流畅，通顺(10 分)	
内容(30 分)	内容完整，涵盖全部写作要点(15 分)	
	内容表述精准，忠于原文(15 分)	
语言描述(30 分)	用词规范、准确，无语法错误(20 分)	
	句式富于变化(10 分)	
其他(20 分)	无雷同状况(10 分)	
	无严重偏题(10 分)	
合　计(100 分)		

Task 4　Distribution Arrangement

知识目标

◆ 学习跨境电子商务中物流配送的内容
◆ 学习跨境电子商务中物流配送的艺术
◆ 学习相关商务沟通的书面英文表达方式

能力目标

◆ 能够准备物流配送材料
◆ 能够顺利开展物流配送活动
◆ 能够处理物流配送的往来信函

Basic Knowledge Review

1. How do you fulfill your online orders?

You can determine if you're able to handle the fulfillment duties on your own or if it's in the best interest of your company to hire a third-party business to take care of fulfilling your online orders. Before you make the right decision, you should weigh the options carefully and consider the following:

(1) Do you have a place for inventory storage?

(2) Do you have the time to handle packaging the shipments as quickly as possible?

(3) Are you good with order and warehouse management?

(4) How much will storage fees be?

(5) Can you meet customer expectations?

(6) Can you offer a privacy policy?

(7) Do you have the time to handle the shipping process and multiple shipping options as quickly as possible?

(8) Are you technologically-savvy and able to be processing orders with fulfillment software?

(9) Can you handle the amount of product fulfillment duties?

(10) Do you want to handle the fulfillment duties for the long term?

If you answered "no" to the questions above, you should hire an order fulfillment company to help you achieve customer satisfaction. If you can afford to outsource e-commerce order management to another company, it might be in your best interest to do so. When you choose to use a third-party logistics company to handle the inventory, packing, shipping, customer service, and other duties for you, the time you save can be used for building up your business.

Consider these aspects when selecting a fulfillment company:

(1) Integrations with your e-commerce platform

(2) Order minimums

(3) The type of businesses they serve

(4) Warehousing and fulfillment costs

(5) How fast they turn around and ship orders

(6) Customer support availability and contact information (phone, live chat, e-mail)

(7) Real time reporting management system

(8) Product order volume capabilities

(9) Type of inventory system software and warehousing capabilities

(10) Network of fulfillment centers, location and warehouse space

By considering all these things and how they relate to your individual business, you can narrow down your fulfillment center options. When you do so, you'll be able to select the best services provider to help your company achieve its multifaceted sales goals.

2. What does an e-commerce fulfillment provider do?

E-commerce fulfillment companies specialize in order processing and delivery for e-commerce businesses. This process includes receiving and storing inventory, processing orders, pick and pack, inserting packing slips, and shipping.

Essentially, they have software that connects to your website, so once and order is placed. They assemble your order into a package, scan it and ship it out. And all this information is communicated back to your website. So, let's say you are selling shampoo. You probably have different sized bottles, and different scents. You can send all those directly to your fulfillment warehouse. Once they get an order, they will pick the size, scent and quantity of your customers order and pack it up into a shipping box.

3. Who are the top 10 world's largest logistics brands?

(1) UPS

United Parcel Service is an American multinational package delivery company and a provider of supply chain management solutions. Services also include a cargo airline, a freight-based trucking operation and retail-based packing and shipping centres. UPS employs approximately 444,000 staff with approximately 240,000 drivers with international package operations delivering to more than 220 countries. Its CEO is David P. Abney and the brand value of the company is $22bn.

(2) FedEx

FedEx Corporation is a US multinational courier delivery services company with its headquarters in Memphis, Tennessee. FedEx offers a complete suite of online services for shipment preparation, package tracking, shipment rates and tools for international shippers and small businesses. The name FedEx is actually an abbreviation of the name of the company's original air division, Federal Express. Brand value, according to Brand Finance, is $18.1bn. The CEO is Frederick W. Smith.

(3) Japan Railways Group

The Japan Railways Group is more commonly known as the JR Group and consists of seven for-profit companies that took over most of the assets and operations of the government-owned Japanese National Railways in 1987. The JR Group has a total route length of about 12,500 miles, of which about half is electrified. Current brand value is thought to be $11.1bn.

(4) DHL

DHL is one of the most recognisable logistics brands across the globe, specialising in international shipping, courier services, road and rail transportation, air and ocean freight, international parcel and express mail services and contract logistics. Founded in the US in 1969, the company had gone global by the late 70s and its current value is thought to be in the region of $10.7bn.

(5) Union Pacific

Union Pacific is a freight hauling railroad that operates 8,500 locomotives consisting of 43 different models, operating over 32,100 route miles. Its system is the second biggest in the US. The company started in back in 1862 when it was called the Union Pacific Rail Road and its current CEO is Lance M. Fritz. The brand value of the company is thought to be $7.8bn.

(6) McLane Company

The McLane Company is of the largest supply chain service leaders providing grocery and foodservice supply chain solutions for convenience stores, drug stores and more in the US. It has one of the largest private fleets in America and delivers around 50,000 products to 110,000 locations and employs around 20,000 employees. The President and CEO of McLane since 1995 is W. Grady Rosier and its brand value is currently $4.8bn.

(7) Poste Italiane

Poste Italiane is an Italian postal service provider. Outside of its standard postal services it also offers various integrated products such as postal savings, communication, logistics and financial services throughout Italy. In 2011, the business acquired UniCredit Medio Credito Centrale for €136mn ($166mn) and in 2016 the Italian government approved the sale of stakes of up to 40% in the company. Its CEO and MD is Matteo Del Fante and its brand value stands at $4.8bn.

(8) CN

With a brand value of $4.4bn, CN is a Canadian-based transportation company that offers integrated services covering rail, intermodal, trucking, freight forwarding, warehousing and distribution. CN has approximately 24,000 railroaders and transports more than $250bn worth of goods annually for an array of business sectors, ranging from resource products to manufactured products to consumer goods, across a rail network of approximately 20,000 route-miles spanning Canada and mid-America.

(9) Deutsche Post

Deutsche Post is part of the Deutsche Post DHL Group and is Europe's leading postal service provider. Deutsche Post delivers mail and parcels in Germany and across the world and provides dialogue marketing and press distribution services as well as corporate communications solutions. Deutsche Post DHL Group employs approximately 510,000 employees in over 220 countries and territories worldwide. The brand value, according to Brand Finance, is $4.2bn.

(10) CSX

CSX is a leading supplier of rail-based freight transportation based out of Jacksonville, Florida. As a railroad, the company operates around 21,000 route miles of track. The company was formed in 1981 by combining the railroads of the former Chessie System, Seaboard Coast Line Industries then the Seaboard System Railroad in 1986. The brand is currently valued at $4bn.

Task Descriptions

杭州红粉服饰有限公司跨境电子商务部运营专员 Lily 在线处理跨境电商平台客户下单前关于物流方面的具体事宜。

Distribution Arrangement

Typical Samples

◇ Sample Ⅰ

Lily 按客户要求安排货物的包装后，给买家发去确认函。下面是关于包装确认函的具体内容。

Dear Sir,

Thanks for the prompt payment on the stockings. We're all set to arrange the shipment right away. According to your request, we will pack each pair of stockings in a plastic bag, 12 pairs in one big bag, and 20 dozen big bags in a carton. Bubble wrap will be used to make sure everything stays in good condition inside the carton. If you agree the goods can be packed as above, we'll ship them later today.

We look forward to your early confirmation.

Best regards,

Pinklife Team

◇ Sample Ⅱ

货物妥投后，Lily 给买家发确认函。下面是确认函的具体内容。

Dear Janey,

We've already prepared your order to be shipped in 2 hours by our shipping agent. Your parcel is being handled by China Post. The tracking number is HJ9837706989CN. It is estimated to reach you within 30 business days. Please note that it may be a little later than normal if there is bad weather for flight delays.

Thanks for your understanding and patience.

Best regards,

Pinklife Team

◇ Sample Ⅲ

货物到达买家所在地，Lily 给买家发去通知函，下面是函电的具体内容。

Dear Linda,

Hello! I appreciate the time you have taken to contact us about your order with tracking number RL7087775676UK and I am happy to assist today.

Here is a link for you to track your package status in transportation: XXXX

I'm glad to inform you that as per the current tracking result your package has arrived in your country on 2020/10/28.

Kindly be advised that when item arrives, it will be processed by customs then hand over to local carrier for final mile delivery. This means your package has arrived at the destination country but you need to wait for clearance before being delivered. Normally, the delivery is expected in coming days.

For delivery related questions, please contact your carrier as below.

CARRIER'S INFO:
Carrier name: USPS
Helpline: 1-800-222-1811

I truly wish you will be satisfied with your purchase and you will confirm your order as soon as you get the item. Of course, if you like the item, please leave us a positive feedback.

Thanks for your business again!

Have a nice day!

Best regards,
Pinklife Team

◇ Sample IV

买家发函询问物流长期不更新是什么原因。下面是函电的具体内容。

Hi, I wonder why the logistics information of the order on Oct. 4th hasn't been updated. I ordered the dress 15 days ago. If the dress is missing, I want a refund. If the dress is on the way, please tell me the exact place where my order is now. If you are on line, please answer me immediately. If I cannot get a reply in 24 hours, I will apply for a refund directly on the platform. Waiting for your massage. Thanks.

◇ **Sample V**

收到买家询问订单物流信息后，Lily 立即联系物流公司调查货物去向，发现货物在运输途中丢失。Lily 将情况汇报公司，经公司同意向买家发函，同意退款。下面是函电的具体内容。

Dear Sam，

We are so sorry to inform you that the logistics company has lost your goods during the delivery. We contact the logistics customer service to investigate the tracking information as soon as we received your letter. Because the peak season is coming, the logistics production is poor.

According to your request, we agree to refund you in 24 hours after you send your refund request on the platform.

Looking forward to your next shopping time.We will give you a 10% discount off for your next order.

Best regards,

Pinklife Team

💞 *Typical Expressions*

1. We offer drop shipping service. You can simply specify the shipping address, and we will deliver the order to your designated address. 我们提供代发服务，只要提供邮寄地址，我们就会将货品发至指定地址。

2. We would check the quality of the product and try our best to make sure you receive it in a satisfactory condition. 发货前我们会检查货物质量，保证您收到的货物完好无损。

3. Your item will be arranged within 24~28 hours then you'll get a courier number. 您的订单将在 24~28 小时内安排发货，届时您可以收到快递单号。

4. Before you check out, make sure to verify or change your delivery address. 在您付款前，一定要核对或者修改您的配送地址。

5. We usually pack each piece of men' shirt in a box and put half a dozen in a carton. 通常每箱含 6 件男衬衫，每件衬衫都有单独包装盒。

6. The wooden case should be not only seaworthy but also strong enough to protect the goods from any damage. 木箱不仅要适于海运，还要足够坚固以保护货物不受任何损害。

7. The package left the seller's facility and is in transit to the carrier. 快件已出库，正在发往快递公司途中。

8. Your package was delivered and the delivery was signed by the buyer. 快件已经成功投递。快

件已签收，收件人是本人。

9. Normally it will take DHL Express about one week to ship the package to the destination. 通常情况下，敦豪速递要一周才能将快递送到目的地。

10. Be patient. The logistics tracking information is often updated after the goods are actually transported. 请耐心等待，物流跟踪信息的更新往往比货物实际运输速度要晚一些。

Composition-Translation

◇ Part Ⅰ　Integrating the following Chinese basic information into English.

亲爱的买家：

　　很抱歉地通知您，由于 2019 新型冠状病毒疫情的影响，阿里旗下的速卖通平台的部分物流以及运输将出现延迟发货的情况，买家收货等待时长需根据实际情况决定。

　　虽然是不可抗力造成订单不可履行，卖家是免责的。但是考虑到您是我们的老客户，而且订单是刚下的。如果您等不及，可以在后台直接取消订单或者申请退款。我们将在 24 小时之内帮您后台处理退款事宜。

　　再次为给您带来不便感到抱歉，欢迎您下次光临购物。谢谢！

您的，
Pinklife Team

◇ Part Ⅱ　Integrating the following English basic information into Chinese.

Dear valued customer,

Hello! I appreciate the time you have taken to contact us about your order with tracking number LO87642006UK and I am happy to assist today.

I regret to inform you, as per the current tracking result, I found the tracking result indicates that your package is undergoing an unusual condition. It was being returned back to us. The possible reason of returning will be written on the package (Incorrect/illegible/incomplete address, expired retention period; the addressee failed to collect the item; the addressee does not reside at the given address; refused to accept by addressee, etc.). I can inform you further once it is returned back to us.

Depending on this unusual situation, please accept my apologies for the inconvenience. I would be willing to resolve this problem by making a full refund or replace one and send to you.

Again, please accept my sincere apologies and let us know which method you prefer to resolve this unfortunate situation.

Have a nice day!

Best regards,
Pinklife Team

Self-Assessment

Evaluate your practice by marking the following each corresponding item and work out the total scores.

考核情境(分值)	考核要求(分值)	得分
行文逻辑(20分)	层次清楚，思路清晰(10分)	
	行文流畅，通顺(10分)	
内容(30分)	内容完整，涵盖全部写作要点(15分)	
	内容表述精准，忠于原文(15分)	
语言描述(30分)	用词规范、准确，无语法错误(20分)	
	句式富于变化(10分)	
其他(20分)	无雷同状况(10分)	
	无严重偏题(10分)	
合　计(100分)		

Task 5　After-Sale Customer Service

知识目标

◆ 学习跨境电子商务中售后客户服务的内容
◆ 学习跨境电子商务中售后客户服务的艺术
◆ 学习相关商务沟通的书面英文表达方式

能力目标

◆ 能够准备售后客户服务材料
◆ 能够顺利开展售后客户服务活动
◆ 能够处理售后客户服务的往来信函

Basic Knowledge Review

1. What does the e-commerce after-sale service involve?

The e-commerce after-sale service aims at guaranteeing positive contact between the buyer and the seller. The seller always needs to be ready to handle product returns and refunds.

Offering a quality after-sale service is just another marketing tool that can win you repeat purchases. After-sale service is an additional cost for e-commerce companies, whether it be in the delivery of returns or product refunds. It impacts the brand perception of a retailer too, as unhappy customers may leave negative reviews following a poor experience.

2. How do you optimize and manage your after-sale service for e-commerce effectively?

The following tips are helpful for you to improve your after-sale service of e-commerce effectively.

(1) Develop a product returns management policy

Your on-site return policy should be clear and available in all the languages of the countries where your website is delivered. For the success of your online business, it's essential to take into account a foolproof return policy. All is to make your customers return easily if they have a problem with the order.

(2) Reduce your online return rate

As we all know, it's impossible to have a 0% e-commerce return rate. Objective and relevant product descriptions are helpful for customers making a purchase decision and help to minimize potential returns. As a sale, its important and effective, if you handle the product return in a positive and timely manner, it can win you more sales in the future. Don't forget to analyze the reasons given by your customers for product returns too for quick ways to optimize.

(3) Facilitate contact between you and your customers

Facilitate contact between you and your customers means your after-sale service is one of the best ways to build online trust with your customers. To create FAQ(Frequently Asked Questions)page to save you countless hours every week answering the same questions. Conducting a customer satisfaction survey can help you win over customers and let them know that you are listening. Another way to win a customer is to offer enough support after the purchase of a product.

Customers don't just buy your product, they are also buying the customer service that you will provide after the sale. A fully-optimized and effective after-sale service can bring you: increased sales, improved customer loyalty, differentiation from the competition and a strong brand identity.

Task Descriptions

杭州红粉服饰有限公司跨境电子商务部运营专员 Lily 处理各大跨境电商平台的买家关于售后反馈问题等事宜。

After-sale Customer Service

Typical Samples

◇ Sample I

Lily 查看 Amazon 后台买家与卖家消息服务(Buyer-Seller Messaging Service on Amazon) 中买家发来的关于商品质量问题要求换货的信函。买家消息留言内容如下：

Message board:

I am unhappy with the quality of the dress on 15 December and I am writing to seek a replacement.

The left sleeve port was open. I noticed this problem as soon as I unpacked it from the box.

The dress is not of acceptable quality and does not match the sample dress described in the store. I would like you to replace it with one of the same quality and finish as the sample and arrange for return of the faulty dress at no cost.

I would like to have this problem fixed quickly please. If I do not hear from you within 24 hours, I will lodge a formal complaint with Amazon Consumer Service.

Yours sincerely,

Rubby

◇ **Sample Ⅱ**

　　Lily 根据买家要求，重新给买家发货。下面是 Lily 给买家的回信。

Respected Sir,

This letter is with reference to your complaint number 2378 regarding the dress you had purchased online from us on 15th December. We are extremely sorry that the articles you had requested were changed when you had got them delivered at your end. We understand how much this had been upsetting for you. You had waited for a period of 14 to 15 days for delivery and the day of delivery the wrong articles were handed over to you by us.

We sincerely apologize to you for not delivering you the right shirts. We have already decided on the next delivery at your place and this time you will not have to pay for the delivery charges taken by us. The requested dress has been packed for shipment at your place and you will soon get an email from our team that tomorrow the dress will be delivered to you.

We are very sorry for the inconvenience caused to you by us. We hope that you will keep purchasing from us.

Yours Sincerely,

Lily

◇ **Sample Ⅲ**

　　Lily 在阿里国际站平台处理了买家的退款要求后，又写了封道歉信给买家根据买家要求退款，下面是 Lily 给买家的回信。

Dear John,

Thanks for reaching out. Satisfying our customers is very important to us and I'm sorry our dress didn't meet your expectations. I fully respect your decision and can only apologize for any problems your business experienced.

We've processed your refund, and you should expect to see the amount credited to your account in about 3 to 5 business days.

If you have any other questions or concerns, just reply to this email, I'll be here to help you in any way I can.

Best,
Pinklife Team

◇ Sample Ⅳ

　　Feedback(Customer Feedback)买家评价是买家在发生真实购买记录之后留下的对卖家服务水平、物流时效和产品信息等信息的评论。买家的评价内容会影响后续买家购买行为。下面是 Lily 给买家发的邀请留评信。

Dear Shelly,

You have made a recent purchase with us, and we thank you for it. Your feedback on the purchase means a lot to us, and we really would like to improve your shopping experience at every chance we get. Please spare a few minutes of your time to help us grow.

Regards，
Pinklife Team

◇ Sample Ⅴ

　　公司针对购物满 50 件的老客户开展促销活动，下面是 Lily 给买家发的促销信。

Dear John,

Congratulations!!

You have completed 50 product purchases from us, and we are ever grateful for your support and trust in us. To make your 50th purchase a special one, we would like to offer you a flat 50% off on specific products during your next shopping.

We are going to make your every purchase a special one, thank you for being with us and helping us grow.

Please use this offer coupon for your next purchase. It is specially designed for you.

Regards,
Pinklife Team

Typical Expressions

1. I got the package, but the item is in a different color. 包裹收到了，但产品颜色不对。

2. I am disappointed with the item. I'll return it because nothing is like the picture. 我对商品很失望，要求退货。产品与图片内容完全不符。

3. Our customers are complaining about the slowness of the delivery. 我们的客户都在投诉货运很慢的问题。

4. The dress is much smaller in size than expected. It's a terrible dress. 衣服尺码比预想的要小得多，真是件糟糕的衣服。

5. The skirt arrived promptly, but the material is thin and starting to fade after two washes. I'm very unhappy with the products. 裙子很快就到了，但是面料很薄，洗两次就褪色了。我对产品很不满意。

6. We made a mistake about the SKU number and sent you the wrong goods. 我们把货号弄错了，因此发错了物品。

7. We'd like to express our appreciation for giving us an opportunity to make it up to you. 谢谢您给我们弥补过失的机会。

8. Please take a look at some other products. We agree to send a replacement. 请看看其他产品。我们同意发一个替换产品。

9. We'll try our best to make up for the loss in accordance with your request. 我们将竭尽全力按您的要求弥补损失。

10. We hope that this unfortunate incident will not affect the relationship between us. 我们希望这一不幸事件不会影响到我们双方之间的关系。

Composition-Translation

◇ Part Ⅰ Integrating the following Chinese basic information into English.

您好，
　　得知发出的裙子破损我们很抱歉。给您带来不便，我深表歉意。根据您的要求，立即给您重新安排发货。发生这种事情真不是我们的本意。希望这件事情不会影响您再次体验我们的新品。

我已经按您的要求退还您的这个包裹的全部费用，包括运费在内共计 10 美元。退款将在 2～3 个工作日打入您的信用卡。您在后台随时可以查看账户退款的状态。

为客户提供方便、高效服务是我们一直追求的目标。出现这种事情，我们真的没想到。希望您能再给我们一次机会，为弥补给您带来的伤害，我们将在您下一次购物的总价中给您 8 折优惠。优惠券已发附件，请收好。

　　　　谢谢！

Pinklife Team

◇ Part Ⅱ　Integrating the following English basic information into Chinese.

Complaint Letter to Amazon about an Item Delivered to the Wrong Address

Dear sirs，

I am writing this letter to let you know about my complaint that, I have been very disappointed with your poor service of delivering my items to the wrong address on the day of April 1.

I ordered a package of dresses on the day of March 27 which was expected to be delivered at my home address.

As the delivery date was March 28, I was eagerly waiting to receive the item. Unfortunately, after three days I had received the news that it had been delivered to the wrong address, though your employees had already confirmed my address via phone call.

Though I tried to contact with your customer care service many times without any delay for a proper delivery, but they were unable to help me till now. I didn't expect this type of careless service from such a well-known organization of yours.

Hence, I am writing this letter to you, to help me with this matter as soon as possible. So, I want a response from your end regarding this matter.

I am looking forward to get your response very soon.

Yours,
John

Self-Assessment

Evaluate your practice by marking the following each corresponding item and work out the total scores.

考核情境(分值)	考核要求(分值)	得分
行文逻辑(20 分)	层次清楚，思路清晰(10 分)	
	行文流畅，通顺(10 分)	
内容(30 分)	内容完整，涵盖全部写作要点(15 分)	
	内容表述精准，忠于原文(15 分)	
语言描述(30 分)	用词规范、准确，无语法错误(20 分)	
	句式富于变化(10 分)	
其他(20 分)	无雷同状况(10 分)	
	无严重偏题(10 分)	
合　计(100 分)		

Task 6　Complaints Settlement

知识目标

◆ 学习跨境电子商务中投诉的内容
◆ 学习跨境电子商务中投诉处理的艺术
◆ 学习相关商务沟通的书面英文表达方式

能力目标

◆ 能够准备投诉处理材料
◆ 能够顺利开展投诉处理活动
◆ 能够处理投诉往来信函

Basic Knowledge Review

1. What are dispute resolutions for cross-border e-commerce platforms?

Generally, there are three dispute resolutions for cross-border e-commerce platforms. They are platform mechanism, alternative dispute resolution and lawsuits.

(1) Platform mechanism

Exporters generally open shops on cross-border e-commerce platforms such as AliExpress, an online marketplace run by Alibaba. Such platforms provide online dispute resolution mechanisms.

These mechanisms clarify the determination of responsibilities, solutions, penalties after any dispute arises, and a large number of disputes are resolved through the mechanism provided by the platform.

(2) Alternative Dispute Resolution (ADR)

In addition to the mechanism provided by the platform, ADR is also a common way for dispute resolution. ADR usually includes negotiation, arbitration, and mediation.

Negotiation is the first choice in cross-border e-commerce dispute resolutions. Because of the potentially high cost and long term for traditional cross-border e-commerce dispute resolution, negotiation is conducive to the rapid settlement of disputes and follow-up cooperation between the two parties. It should be noted that mediation settlement agreements are currently often regarded as contracts in China, and thus cannot be mandatorily enforced.

Whereas, due to the high cost, arbitration may not be applicable to cross-border e-commerce dispute resolution with small and medium-sized amounts in controversy.

Therefore, pursuant to the author's investigation, 60% of the enterprises stated that they didn't resort to the ADR methods other than negotiation.

(3) Lawsuits that have to be taken as a last resort
Cross-border e-commerce disputes will be resolved by litigation only when the platform mechanism and ADR are unable to resolve the disputes.

Cross-border e-commerce disputes are often foreign-related disputes. Take intellectual property disputes as an example, many foreign right holders will file lawsuits with foreign courts. Some cross-border e-commerce enterprises in Yiwu said that they had received subpoenas from foreign courts, but the high cost of defending legal proceedings overseas made them prohibitive.

The costs for cross-border litigation are often high and may exceed the value of goods, so more than 80% of companies said that in such lawsuits, they could only allow foreign courts to freeze their platform accounts and seize the relevant commodities.

2. Under what circumstances can you open a dispute on AliExpress?
✓   Faulty product or item not as described.
✓   Slightly imperfect product.
✓   Incomplete order.
✓   The order didn't arrive.
✓   The product is fake.
✓   You get something you didn't order.
✓   Order delivered but not received.
✓   The shipping method is not correct.

Task Descriptions

杭州红粉服饰有限公司跨境电子商务部运营专员 Lily 与某跨境电商平台协商处理关于产品侵权、客户投诉、账号暂封等事宜。

Complaints Settlement

Typical Samples

◇ Sample Ⅰ

当亚马逊平台判定账户发生关联，会给卖家注册邮箱发账户关联警告信。下面就是 Lily 收到的来自亚马逊客服的账户关联警告信原文。

Hello from Amazon.com

We are writing because it has come to our attention that you are operating multiple merchant accounts.

This activity is in violation of policies, which state that "operating and maintaining multiple seller accounts is prohibited."

In accordance with this policy, please cancel all listings on your xxxx account, and limit your selling activities to your xxxx2 account.

Please note that failure to comply with this request may result in the blocking of all of your selling accounts.

We appreciate your cooperation in this important matter.

Regards,

Seller Performance Team

Amazon.com

◇ Sample Ⅱ

当亚马逊平台判定卖家发生侵权行为时，会给卖家注册邮箱发侵权通知函。下面就是 Lily 收到的来自亚马逊客服的侵权通知函原文。

Hello from Amazon,

We are writing to inform you that your disbursements from your seller account have been placed on hold. We took this action in accordance with a Temporary Restraining Order issued by a federal court. Items that infringe another party's copyright, patent, trademark, design right, database right, or other intellectual property or other proprietary right are prohibited. For more information on this policy, search on "Prohibited Content" in seller help.

To resolve this dispute, we suggest that you contact the rights owner directly.

If you resolve this matter with the rights owner, please advise them to contact us at notice-dispute@amazon.com to withdraw their complaint. We ask that you refrain posting items that infringe the intellectual property rights of this rights owner until you have resolved this matter.

Failure to comply with our policies may result in the removal of your Amazon.com selling privileges.

We appreciate your cooperation and thank you for selling on Amazon.com.

Sincerely,
Seller Performance Team
Amazon.com

◇ Sample III

Lily 收到平台通知函后马上向运营总监报告，得到公司许可后立即向 Amazon 平台客服发电子邮件进行侵权申诉，下面是申诉原文。

Dear sir/madam,

Thanks for your patience about our issue. We extremely apologize for the inconvenience to you.

We got to recognize that we made a mistake and may infringe the intellectual property rights of others. When we received your email about the issue that we have created remove order of our FBA stock to prevent the similar complaint. Now we have send an email to amazonsupport@XXX.com at 6 pm, and plead them to withdraw the complaint and we have to make a commitment to them:

We will never sell these products again in the future and to our behavior made the earnest words of apology, and we also Cc the email to Amazon, but we haven't received any responded up until now. We will keep our eyes on this issue and I hope that we can find an amicable solution to this issue through the consultation process.

We take the following measures after we got your email:

(1) Since we got the Warning Notice of Intellectual Property Rights Infringement about our ASIN: kkl110 may infringe the intellectual property rights of others, we have removed all the inventory stop selling at first time, now we have deleted this listing now and we no longer sell it again.

(2) We have contacted the rights owner for retracting the complaints and made the earnest words of apology, but we haven't received any responded up until now, we will closely watch the reply.

About this issue, we have taken the following steps to help us to resolve the issue and prevent similar complaints.

(1) We have checked all the listing detail from title, image, description and bullet point to ensure have no information in misunderstanding. And we are firmly deleting all the listings that could show any signs of conflicts with intellectual property rights.

(2) We have established the professional handling team to take care of listing 60% match on description, images, bullet points, & search terms, etc. Never make more mistakes of the Product Design or Brand.

(3) We have organized our own research department to make sure we can sell our own products with our own design & model instead of purchasing from our supplier, to prevent any possible infringement problems on Trademarks & Packages & Design.

(4) All of the email or complaint answered and resolve under 24 hours. Provide a good after-sale service.

Hope you can give me a chance again. If there's any other information you need, please feel free to contact us.

Yours sincerely,
Lily

◇ **Sample Ⅳ**

    Lily 发现由于退款率太高导致 Wish 平台封号，马上向运营总监报告，得到公司许可后立即向 Wish 平台客服发电子邮件进行申诉，下面是申诉原文。

Dear Seller Support,

We are really sorry for the high refund ratio right now.

We check all the orders that be refunded and find that almost all the order had been refunded for the long shipping time. It is known that we offer all the orders with tracking numbers and hope to offer a better shipping service for us, resulting in that the shipping time had been delayed.

Based on the situation, we will change the shipping services and use "Wish Express" next year. Also in order to show our sincerity, our company plan to list the product with our USA brand next year. And we will set up a dedicated team for Wish to offer customer a best service.

Please help us re-activate our account. We really want to work with Wish! Thanks for your help.

Yours,

Lily

◇ **Sample V**

Lily 发现由于商品质量及物流太慢导致 Wish 平台封号，马上向运营总监报告，得到公司许可后立即向 Wish 平台客服发电子邮件进行申诉，下面是申诉原文。

Dear Seller Support,

I feel shocked and really sorry about this, we always provide best products and service for customers.

We are so contributed on the products and didn't notice the vender is irresponsible for his products that is why now we face this situation.

We have changed the vender and promise we will never make this happen again.

Now we have the best vendors and the best products also we are planning to use Wish Express to make sure we can provide the fastest and safe express service to customers.

We have completed plenty of orders on Wish. I think customers also feel good to use it. Most of our products were designed and developed by ourselves, we have the best price and they are unique on Wish.

I believe we will both have a nice experiment by cooperating with each other. Hope you can give us a chance to prove ourselves.

Thank you very much.

Best regards,

Yours,

Lily

🐝 *Typical Expressions*

1. The product does not infringe the intellectual property rights of the XXX brand. 该产品没有侵犯 XXX 品牌的知识产权。

The product does not infringe the intellectual property rights of any brand.该产品没有侵犯任何品牌的知识产权。

2. This product is not XXX banned by Wish. 该产品不是 Wish 禁售的 XXX。

3. The title, description, label, picture, color, and size of the product are not associated with XXX. 该产品的标题、描述、标签、图片、颜色、尺寸都与 XXX 没有任何关联。

The title, description, label, picture, color, and size of this product are not associated with any brand. 该产品的标题、描述、标签、图片、颜色、尺寸与任何品牌都没有任何关联。

4. The product does not have vague information. 该产品不存在模糊的信息。

5. This product does not contain sexually explicit content. 该产品不含有露骨的淫秽内容。

6. This product will be able to help babies and parents around the world. My shop is going to use a lot of money to do PB for this product. Please re-determine. 该产品将帮到全世界的婴儿以及父母，我的小店准备用大量的钱为该产品做自有品牌，请平台重新判定。

7. This product is just an ordinary product, it has been misjudged, please re-judge. 该产品只是一件普通的产品，它被误判了，请重新判决。

📖 *Composition-Translation*

◇ **Part Ⅰ Integrating the following Chinese basic information into English.**

亲爱的亚马逊，

　　我们之间已经产生了误会，关于"粉红生活(Pinklife)"商店销售假冒/知识产权/商标侵权投诉事件是一个误会。我们并没有提交申诉材料。显然一个名为"蓝色(Blueety)"的实体正在使用 Sun 商标并向亚马逊提交了虚假报告。

　　作为卖方"粉红生活(Pinklife)"并没有侵犯任何知识产权，也没有试图出售 Sun 的任何假冒商品。

　　如果能立即撤回上述两项投诉，并将其从亚马逊的"粉红生活(Pinklife)"账户中删除，我们将不胜感激并增加在本产品中自有品牌的投入。

如果有必要需要我们提供除上述内容以外的任何额外信息，请立即通知我们。谢谢您的支持！

您的

Lily

◇ Part Ⅱ Integrating the following English basic information into Chinese.

Hi there,

From my observations of the information you provided, it looks like you have submitted everything we would ask for to appeal an infringement complaint.

My only concern is that you are the individual who has complaints against your listings, and alongside your appeal, you are submitting a 'letter from the manufacturer' who is speaking on your behalf.

The retraction of a complaint has to come from the rights owner directly, or the individual/company that has registered their brand/provided proof of their rights to be administered on the Amazon store.

If you buy your items directly from the manufacturer, the communication will have to come from them directly to either Amazon or the rights owner who filed the complaint against you on Amazon.

At that point, there will have to be a future determination if you will be allowed to utilize the intellectual property that was infringed upon by the complaining parties involved.

In regards to your invoice(s), I cannot speak to them since I do not have the document to review. I would suggest reviewing the information to verify if it meets the following requirements:

- Supplier issued in the last 365 days for the ASINs listed below
- These documents should reflect your sales volume during the last 365 days.
- Please include contact information for your supplier, including name, phone number, address, and website. We will maintain the confidentiality of your supplier contact information. **
- You may remove pricing information, but the rest of the document must be visible. For

ease of our review, you may highlight or circle the ASIN(s) under review.

Best,

Arthur

Self-Assessment

Evaluate your practice by marking the following each corresponding item and work out the total scores.

考核情境(分值)	考核要求(分值)	得分
行文逻辑(20 分)	层次清楚，思路清晰(10 分)	
	行文流畅，通顺(10 分)	
内容(30 分)	内容完整，涵盖全部写作要点(15 分)	
	内容表述精准，忠于原文(15 分)	
语言描述(30 分)	用词规范、准确，无语法错误(20 分)	
	句式富于变化(10 分)	
其他(20 分)	无雷同状况(10 分)	
	无严重偏题(10 分)	
合　计(100 分)		

Appendix 1   Ali-express 平台中英文对照

Apparel & Accessories 服装/服饰配件

Apparel Accessories 服饰配饰(男/女/儿童配件，婴儿配饰发到婴儿服装)

Arm Warmers 手臂套

Belts 腰带/皮带

Cummerbunds 腰封

Earmuffs 耳罩

Eyewear & Accessories 眼镜及配件

Chains & Lanyards 眼镜链

Eyeglasses Frames 眼镜架

Eyeglasses Lenses 眼镜片

Eyewear Cases & Bags 眼镜盒/眼镜袋

Lens Clothes 擦镜布

Reading Glasses 阅读镜

Sunglasses 太阳镜

Gloves & Mittens 手套

Handkerchiefs 手帕

Hats & Caps 帽子

Baseball Caps 棒球帽

Berets 贝雷帽

Bomber Hats 雷锋帽/护耳帽

Bucket Hats 渔夫帽

Cowboy Hats 牛仔帽

Fedoras 礼帽/桶帽

Military Hats 军帽

Newsboy Caps 报童帽

Skullies & Beanies 无檐便帽/套头帽/蒙面帽

Sun Hats 太阳帽

Visors 鸭舌帽

Headwear 头饰

Scarf, Hat & Glove Sets 多件套(围巾帽子手套)

Scarves & Wraps 围巾/披肩

Suspenders 吊裤带

Ties 领带

Costumes & Accessories 扮演服及配件

Boas 围巾

Gloves & Handwear 手套

Hats 帽子

Headwear 头饰

Masks & Eyewear 面具和眼镜

Sashes 腰带

Stockings, Tights & Socks 长筒袜，连裤袜和短袜

Ties 领带

Weapons 武器

Costume Kits 扮演服饰套装

Costumes 扮演服装

Lolita Dresses 洛丽塔裙装

Mascot 卡通人偶

Scary Costumes 恐怖扮演服

Sexy Costumes 性感扮演服

Zentai 紧身衣

Men's Clothing 男装

Active Tracksuits 休闲运动服套装

Coats & Jackets 外套/大衣

Down & Parkas 羽绒服/棉服

Jackets 夹克

Leather & Suede 皮衣

Trench 风衣

Vests 外套背心

Wool & Blends 呢子大衣

Hoodies & Sweatshirts 卫衣帽衫

Jeans 牛仔裤

Men's Sets 男士套装

Men's Sleep & Lounge 睡衣/家居服

Onesies 连体睡衣

Pajama Sets 睡衣睡裤套装

Robe Sets 睡袍睡裤套装

Robes 睡袍

Sleep Bottoms 睡裤

Sleep Tops 睡衣

Pants 长裤

Shirts 衬衫

Shorts 短裤(外穿)

Socks 袜子

Suits & Blazer 西服

Blazers 休闲西服

Suit Jackets 正装西服外套

Suit Pants 西装长裤

Suits 西服套装

Vests 西装马甲

Sweaters 毛衣

Swimwears 泳衣/沙滩服

Board Shorts 沙滩短裤

Briefs 三角泳裤

Rash Guards 冲浪服/沙滩服上衣

Trunks 平角泳裤

Tops & Tees 上衣，T恤

Polo Shirts Polo 衫

T-Shirts T恤

Tank Tops 背心

Underwears 内衣

Bikinis 比基尼裤

Boxers 平脚裤

Briefs 三角裤

G-Strings & Thongs 丁字裤

Long Johns 长内衣裤套装

Shapers 塑身衣

Undershirts 内穿汗衫

Exotic Apparel 情趣服装

Babydolls & Chemises 性感睡裙(短、露、透)

Bras 文胸

Costumes 情趣制服

Dresses 连衣裙

Exotic Dresses 情趣连衣裙

Exotic Pants 性感裤

Exotic Tanks 性感背心

Lingerie Sets 性感内衣套装

Panties & Briefs 内裤

Socks & Hosiery 袜子

Teddies & Bodysuit 连体衣/全身连体袜

Stage & Dance Wear 舞台表演服和舞蹈服

Ballet 芭蕾舞服

Ballroom 摩登舞服

Belly Dancing 肚皮舞服

Chinese Folk Dance 中国民族舞服装

Flamenco 弗拉门科舞服

Latin 拉丁舞服

Work Wear & Uniforms 工作服/制服

Food Service 厨师服装

Medical 医学/实验室服装

Military 军装

School Uniforms 校服

Scouting Uniforms 童子军制服/其它服装/
运动服饰

Armwarmers 手臂套

Glasses 运动眼镜

Gloves & Mittens 运动手套/运动连指手套

Hats & Caps 运动帽子

Legwarmers 护脚套

Scarves & Wraps 运动围巾，运动披肩

Socks 运动袜

Clothing Sets 运动套装

Coats & Jackets 运动外套，运动夹克

Down & Parkas 运动羽绒服，运动棉服

Dresses 运动连衣裙

Hoodies & Sweatshirts 运动卫衣帽衫

Jerseys 球服，运动衫

Pants 运动裤

Shirts 运动衬衫，运动 POLO 衫

Skirts 运动半身裙

Sweaters　运动毛衣

T-Shirts　运动 t 恤

Tanks & Camis　运动背心，运动吊带

Vests　运动马甲

Special Occasion Dresses　特殊场合服装

Celebrity Dresses　明星礼服

Cocktail Dresses　鸡尾酒会礼服裙

Communion Dresses　圣餐礼服

Evening Dresses　晚礼服

Graduation Dresses　毕业礼服

Homecoming Dresses　同学会礼服

Party Dresses　派对礼服

Prom Dresses　舞会礼服

Quinceanera Dresses　成人礼礼服

Wedding Accessories　婚庆配饰

Bridal Gloves　新娘手套

Bridal Hairpins　新娘头花

Bridal Hats　新娘帽子

Bridal Umbrella　新娘花伞

Bridal Veils　新娘头纱

Petticoats衬裙

Rose Petals　玫瑰花瓣

Wedding Bouquet　新娘捧花

Wedding Jackets / Wrap　婚纱披肩

Wedding Dresses　婚纱

Wedding Party Dress　婚宴礼服

Boys' Attire　男童服装

Wedding Party Dress　伴娘礼服

Flower Girl Dresses　花童服

Groom Wear　新郎礼服

Mother of the Bride Dresses　妈妈服/女装

Active Tracksuits　休闲运动服套装(非专业运动服)

Blazer & Suits　休闲西装/正装西服

Blazer & SuitsBlazers　单件西装

Blazer & Suits Dress Suits　连衣裙＋西服套装

Blazer & SuitsPant Suits　长裤＋西服套装

Blazer & Suits Skirt Suits　半身裙＋西服套装

Blouses & Shirts　雪纺衫/衬衫

Coats & Jackets　外套/大衣

Down & Parkas　羽绒服/棉服

Fur & Faux Fur　假皮草及混合皮草

Jackets　夹克/短外套

Leather & Suede　皮衣

Real Fur　真皮草

Trench　风衣

Vests　外套背心/马甲

Wool & Blends　呢子大衣

Dresses　连衣裙

Hoodies & Sweatshirts　卫衣帽衫

Intimates　贴身衣物

Bra & Brief Sets　文胸套装

Bras　文胸

Busters & Corsets　紧身胸衣

Camisoles & Tanks　吊带/背心(仅内穿)

Garters　吊袜带

Intimates Accessories　内衣配件

Long Johns　长内衣裤

Panties　女士内裤

Safety Short Pants　安全裤

Shapers　塑身美体内衣

Slips　衬裙

Tube Tops　裹胸/抹胸

Jeans　牛仔裤

Jumpsuits & Rompers　连体衣/背带裤

Leggings　打底裤

Pants & Capris　长裤/九分七分五分裤

Shorts　短裤

Skirts　半身裙

Socks & Hosiery　袜子

Leg Warmers　暖脚套

Peds & Liners　袜垫

Sock Slippers　船袜

Socks　短袜

Stockings　长筒袜

Tights　连裤袜

Sweaters 毛衣

Swimwears 泳衣/沙滩服

Bikinis Set 比基尼套装

Board Shorts 沙滩短裤

Cover-Ups 沙滩裙/沙滩上衣/披纱

One Pieces 连体泳衣

Rash Guards 冲浪服

Tankinis Set 分体泳衣套装

Two-Piece Separates 单件泳衣上装/单件泳裤

Tops & Tees 上衣，T 恤

Polo Shirts Polo 衫

Tanks & Camis 背心，吊带

Women's Sets 女士套装

Women's Sleep & Lounge 睡衣，家居服

Nightgowns & Sleepshirts 普通睡裙

Onesies 连体睡衣

Pajama Sets 睡衣裤套装

Robe & Gown Sets 睡袍睡裙套装

Robes 睡袍

Sleep Bottoms 睡裤

Sleep Tops 睡衣

Africa Clothing 非洲服装

Asia & Pacific Islands Clothing 亚洲及太平洋群岛服装

India & Pakistan Clothing 印巴服装

Islamic Clothing 中东 / 伊斯兰教服

Muslim Fashion 穆斯林服饰

Abaya 女士长袍

Breast Feeding 哺乳装

Dresses 裙装

Jubba Thobe 男士穆斯林长袍

Kebaya 可芭雅

Men's Muslim T-Shirts 男士穆斯林 T 恤

Men's Muslim Shirts 男士穆斯林衬衣

Muslim Sets 穆斯林套装

Women's Bottoms 女下装

Women's Hijabs 女士头巾

Women's Outwear 女外套

Women's Prayer Garment 祷告装

Women's Swimming Suit 女士泳装

Women's Tops 女上衣

Traditional Chinese Clothing 传统中国服装

Bottoms 中式裤子

Cheongsams 旗袍

Robe & Gown 长袍

Sets 中式服装套装

Tops 中式上衣

Automobiles & Motorcycles 汽车、摩托车

Bluetooth Car Kit 车载蓝牙免持听筒套件

Cables, Adapters & Sockets 电缆/适配器/插座

FM Transmitters 调频发射机

Remote Controls 车载遥控器

Speaker Boxes 扬声器机箱

Speaker Mounts 扬声器格板

Subwoofer Boxes and Enclosures 低音炮机箱/外壳

TV-Tuners 电视调谐器

Alarm Systems & Security 警报系统/安全用品

Amplifiers 汽车功放

Audio 音频播放器

Car CD Player 车载 CD 播放器

Car Cassette Player 车载卡带播放器

Car MP3 Player 车载 MP3 播放器

Car Air Purifiers 车内空气净化器

Car Electrical Appliances 车载电器

Heating & Fans 取暖/风扇

Refrigerators 冰箱

Vacuum Cleaner 车用吸尘器

Car PC 车载电脑

Car Radios 汽车收音机

Car Video Players 视频播放器

Car DVD 车载 DVD

Car MP4, MP5 车载 MP4/MP5

Car Monitors 车载显示器

DVR/Camera 行车记录仪/车载录像机

Equalizers 均衡器

Mobile Radio 车用对讲机

Parking Sensors　倒车雷达

Radar Detectors　测速雷达(仅有测速功能)

Rear View Camera　后视摄像头

Speakers　扬声器

Subwoofers　低音音箱

TV Receiver for Car　车载数字电视盒

Convertible Accessoires　敞篷车附件

Exterior Accessories　车外饰/防护

Awnings & Shelters　晴雨挡

Bumpers　保险杠

Car Covers　车衣

Car Stickers　车贴

Car Tax Disc Holders　年检贴

Chromium Styling　高光镀铬装饰

Emblems　车标

License Plate　汽车牌照架

Protective Frames　防护架

Roof Racks & Boxes　车顶行李架

Window Foils & Solar Protection　防爆膜/太阳膜

GPS Accessories GPS 配件

GPS Receiver & Antenna　接收器和天线

GPS Tracker GPS 跟踪器

Marine GPS　船用 GPS

Motorcycle GPS　摩托车 GPS

Sport & Handheld GPS　手持/户外 GPS(非车上使用)

Vehicle GPS　车载 GPS 设备

Interior Accessories　内饰/汽车用品

Air Freshener　空气清新用品

Animal Transportation　动物运输

Anti-Slip Mat　防滑垫(固定手机等设备)

Armrests　扶手

Auto Fastener & Clip　车用夹子、扣件

Car Chargers　车载充电器

Cigarette Lighter　点烟器

Fascias　音响面板

Floor Mats　脚垫

Folding Luggage Cart　车载行李车

Gear Shift Collars　排挡套

Gear Shift Knob　排档头

Glasses Case　眼镜架

Handbrake Grips　手刹套

Handbrakes　手刹头

Interior Mirrors　车内镜

Interior Mouldings　车内装饰条

Key Case for Car　车钥匙改装壳

Key Rings　钥匙环

Mounts & Holder　车载支架

DVR Holders　行车记录仪支架

Drinks Holders　饮料架

GPS Stand GPS 支架

Laptop Stand　笔记本支架

Phone Holders　手机座

Roll-over Bars & Cages　防滚架

Shelves　杂物架

Tablet Stand　平板支架

Universal Car Bracket　通用支架

Nets　杂物袋

Ornaments　车用挂饰/摆件

Parking Assistance　停车辅助(非电子类产品)

Rear Racks & Accessories　后备箱储物/支架

Seat Belts & Padding　安全带/安全带套

Seat Covers　座套/坐垫

Seat Supports　车用靠垫

Sleepy Reminder for Car　防瞌睡提醒器

Steering Covers　方向盘套

Stowing Tidying　收纳整理

Tensioning Belts　张力绳

Motocycle Accessories & Parts 摩托车配件

Care　护理用品

Decals & Stickers　贴纸

Helmet Headset　头盔对讲机

Lubricants　润滑剂

Motocycle Covers　摩托车衣

Motorcycle Ramps　摩托车斜坡道

Paints & Sprays　油漆/喷雾

Safety & Breakdown Assistance 安全与故障

援助

Theft Protection 防盗保护

Tilts & Protective Sheets 车盖/保护膜

Trailers & Trailer Couplings 拖车/挂车接头

Bumpers & Chassis 摩托车保险杠/底盘

Drive & Gears 摩托车传动系统/齿轮

Electrical System 电气系统

Motorcycle Battery 摩托车蓄电池

Motorcycle Motor 摩托车电机

Motorcycle Starter 摩托车起动机

Motorcycle Switches 摩托车转换器开关

Engines & Engine Parts 摩托车发动机/部件

Exhaust & Exhaust Systems 摩托车排气系统

Filters 过滤器

Air Filters & Systems 空气滤清器

Oil Filters 机油滤清器

Frames & Fittings 摩托车框架/配件

Covers & Ornamental Mouldings 机壳/装饰
框条

Engine Guard 引擎保险盖

Falling Protection 落地保护

Fan Cover 风扇罩

Foot Rests 脚踏板

Full Fairing Kits 整流罩套件

Grips 握把

Handlebar 手把

Headlight Bracket 大灯支架

Locks & Latches 锁/插销

Registration Plate Holder 车牌架

Seats & Benches 座椅/长凳

Side Lining 边衬

Side Mirrors & Accessories 侧镜/配件

Stands 停车架

Windscreens & Wind Deflectors 挡风玻璃/
导风板

Fuel Supply 摩托车燃料系统

Ignition 摩托车点火装置

Instruments 摩托车仪表

Lighting 摩托车照明

Motorbike Brakes 摩托车刹车

Protective Gears 摩托车防护用具

Bags & Luggage 摩托车包

Boots For Motorcycle 摩托车防护靴

Combinations 防护套装

Lasses 眼镜/风镜

Gloves 手套

Helmets 头盔

Jackets 夹克

Motorcycle Protective kneepad 防护护膝

Motorcycle Face Mask 摩托车保暖面罩

Motorcycle Luggage Net 摩托车行李网

Pants 长裤

Shirts & Tops 衬衫/T恤

Shorts 短裤

Seat Covers 摩托车坐垫

Wheels & Rims 摩托车车轮/轮圈

RV Parts & Accessories 房车配件和附件

Replacement Parts 专业零配件

Aerials 天线

Air Intakes 进气系统

Air-conditioning Installation 空调设备

Aluminum Oil Cooler 铝板式油冷器

Batteries & Accessories 汽车电池及附件

Boot Flaps 后备箱盖

Brakes 刹车系统

Car Lights 车灯

Car Light Source 车灯光源

External Lights 车外灯

Interior Lights 车内灯具

Car Locks 汽车锁

Chassis Components 底盘零件

Clutch 离合器

Door Handles 门把手

Drive & Transmission 驱动/传动

Engine 引擎

Engine Bonnets 发动机罩

Exhaust & Exhaust Systems 排气系统

Frequency-separating filters 变频分离过滤器

Front & Radiator Grills　前包围/散热器护栏

Front Skirt　前裙板

Fuel Injector　喷油嘴

Fuel Tanks　油箱

Fuses　电路保险

Gears　排挡杆

Ignition　点火系统

Instruments　仪表

Lamp Hoods　车灯保护

Mirror & Covers　车外后视镜/外壳

Mudguards　挡泥板

Multi-tone & Claxon Horns　车喇叭

Nuts & Bolts　螺钉帽/螺栓

Pedals　脚踏板

Petrol Cans　备用油箱

Racing Grills　中网

Reflective Strips　反光条

Seats, Benches & Accessories　汽车座椅

Spoilers　阻流板

Steering Wheels & Steering Wheel Hubs　方向盘(零部件)

Styling Mouldings　防撞胶/装饰条

Suspension　悬挂

Switches & Relays　开关/继电器

Tank Covers　油箱盖

Vehicle Stabilizer Shaft　平衡杆胶套

Wheels, Rims & Accessories　轮胎/配件

Window Lever & Window Winding Handles　升窗系统/曲柄

Windscreen Wipers　雨刷器

Roadway Safety　交通安全

Alcohol Tester　酒精检测仪

Convex Mirror　安全凸面镜

High Visibility Jackets　反光背心

Ice Scraper　除雪铲

Other Roadway Products　其他交通安全产品

Reflective Material　反光材料

Reflective Safety Clothing　反光安全服

Road Stud　道钉

Snow Chains　防滑链

Speed Bump　道路减速设施

Traffic Barrier　交通护栏

Traffic Light　交通警示灯

Traffic Signal　交通安全标志

Warning Triangles　三角警示牌

Tools, Maintenance & Care　汽车工具/维修/保养

Additives　添加剂

Car Jacks　千斤顶

Car Washer　汽车清洗器

Cockpit Care　驾驶舱保养

Code Readers & Scan Tools　读码器/扫描工具

Diagnostic Tools　诊断工具

Emergency Hammer　救生锤

Engine Care　发动机保养

Fillers, Adhesives & Sealants　密封条/黏合剂

Frost Protection　防冻液

Fuel Saver　节油用品

Grease Guns　注油枪

Ice Protection Foil　防冻膜

Inflatable Pump　充气泵

Leather & Upholstery Cleaner　皮革/地毯清洁

Lubricants　润滑剂

Paint Care　漆面保养

Plastic & Rubber Care　塑料/橡胶制品保养

Rim Care　轮毂保养

Rust Converter, Rust Remover & Rust Prevention　防锈/除锈

Sponges, Cloths & Brushes　海绵/抹布/刷子

Tire Repair Tools　车胎维修工具

Towing Bars　拖车杆

Towing Ropes　拖车绳

Tyre Gloss　轮胎上光

Window Cleaning　车窗清洁

Transporting & Storage　运输和储藏

Trailer　拖车

Trailer Couplings & Accessories 拖车链接/
　附件

Beauty & Health 美容健康

Bath & Shower 沐浴用品

Bath 泡澡用品

Bathing Accessories 沐浴工具

Scrubs & Bodys Treatments 身体磨砂/修复
Sets 套装

Shower Gels 沐浴乳

Shower Oils 沐浴油

Soap 香皂

Fragrances & Deodorants
　香水/除臭芳香用品

Antiperspirants 除臭

Deodorants 止汗

Perfume 香水

Hair Care & Styling 头发护理/造型

Conditioners 护发素

Hair & Scalp Treatments 头发头皮修复

Hair Care Sets 护发套装

Hair Color 染发剂

Hair Loss Products 防脱发

Hair Perms & Texturizers 烫发剂

Hair Relaxers 直发膏

Shampoos 洗发水

Styling Products 造型用品

Braid Maintenance 编发造型用品

Curl Enhancers 弹力素

Hair Sprays 造型喷雾

Mousses & Foams 摩丝

Pomades & Waxes 发蜡

Styling Cream 发膏

Styling Gel 造型啫喱

Styling Lotion 造型乳液

Styling Tools 造型工具

Applicator Bottles 洗头瓶

Braiders 编发工具

Brushes 扁梳/圆梳

Cap 帽子

Combs 梳子

Curling Irons 卷发棒/卷发器

Diffusers 烘发筒

Foil 箔纸

Hair Clips 发夹

Hair Color Mixing Bowls 调色碗

Hair Coloring Sets 调色套装(工具类)

Hair Dryers 吹风机

Hair Pins 发针(造型辅助，非装饰性)

Hair Rollers 造型发卷

Hair Scissors 美发剪刀

Hair Trimmers 头发修剪器

Straightening Irons 直发夹板

Styling Accessories 造型配件

Wrap 围巾

Hair Extensions & Wigs 接发/假发

Hair Extensions 接发

Hair Pieces 发片

Bangs 刘海

Chignons 假发髻

Ponytails 马尾

Toupees 顶髻

Wig 头套

Full Wigs 全头套

Half Wigs 半头套或 3/4 头套

Lace Closure 蕾丝小头

Lace Frontal 蕾丝前头

Clips 接发卡子

Color Rings 色板

Connectors 接发器

Glue Sticks 胶棒

Hairnets 发网

Hook Needle 钩针

Lace cap 蕾丝网帽

Links, Rings & Tubes 接发环

Loop brushes 梳子

Pliers 接发钳

Wig Stands 假发支架

Clip-In Hair Extensions 卡子发束

Full Head Set　整头

One Piece　单片

Closure　小头

Fusion Hair Extensions　挂胶发束

Hair Weaving　织发

Loop Micro Ring Hair Extensions　拉环接发

Skin Weft Hair Extensions　皮条发

Toupees　顶髻/男子假发

Wigs　整顶假发

Human Hair　真人发

Hair Weft & Closure　发帘和配件

Health Care　健康保健

Braces & Supports　护具

Breast Enhancement Cream　丰胸膏

Breast Form　义乳

Ear Care　耳部护理

Feminine Hygiene Product　女生卫生护理

Household Health Monitors　家用健康监测器

Blood Glucose　血糖仪

Blood Pressure　血压计

Body Fat Monitors　脂肪测量仪

Pedometer　计步器

Scale　体重秤

Masks　口罩

Massage & Relaxation　按摩

Pill Cases & Splitters　药盒/分药器

Sleep & Snoring　睡眠/止鼾

Slimming Creams　减肥霜

Thermometers　温度计

Makeup　彩妆

Body　身体彩妆

Body Glitter　闪光粉

Body Paint　身体彩绘

Concealer　遮瑕

Foundation　身体粉底

Eyes　眼部彩妆

Concealer & Base　底妆/遮瑕

Eye Shadow　眼影

Eye Shadow & Liner Combination　眼影眼线笔

Eyebrow Enhancers　眉粉/眉笔/眉膏

Eyelash Growth Treatments　睫毛增长液

Eyeliner　眼线

Glitter & Shimmer　闪粉

Mascara　睫毛膏

Face　脸部彩妆

BB & CC Creams BB 霜&CC 霜

Blush　胭脂

Bronzers & Highlighters　高光/阴影

Concealer　遮瑕

Foundation　底妆

Powder　粉饼/蜜粉/散粉

Primer　隔离/妆前/打底

Lips　唇部彩妆

Lip Balm　润唇膏

Lip Gloss　唇彩

Lip Liner　唇线笔

Lipstick　口红

Makeup Remover　卸妆产品

Makeup Sets　彩妆套装

Makeup Tools & Accessories　化妆工具/附件

Cosmetic Puff　化妆扑

Eye Shadow Applicator　眼影棒

Eyebrow Stencils　眉模具

Eyebrow Trimmer　修眉刀

Eyebrow Tweezers　眉钳

Eyelash Curler睫毛夹

Eyelash Glue　睫毛胶

Eyelid Tools　双眼皮工具

False Eyelashes　假睫毛

Makeup Brushes & Tools　化妆刷

Makeup Scissors　化妆剪

Makeup Tool Kits　彩妆用具套件

Nail Art & Tools　美甲用品及修甲工具

Nail Art　美甲艺术

Acrylic Powders & Liquids　水晶粉

Base Coat　底油

False Nails　假指甲

Multi-Use Top & Base Coat　多功能底油盖油

Nail Gel 指甲胶

Nail Glitter 美甲闪粉

Nail Polish 彩色指甲油

Rhinestones & Decorations 水钻/装饰物

Sculpture Powder 雕花粉

Stickers & Decals 指甲贴

Top Coat 盖油

Nail Polish Remover 卸甲相关

Nail Tools 修甲工具

Buffers 抛光块

Callus Shavers 斜口死皮指甲刀

Callus Stones 浮石

Clippers & Trimmers 指甲刀

Cuticle Pushers 死皮推

Cuticle Scissors 死皮剪

Dotting Tools 点笔

Electric Manicure Drills & Accessories 打孔器

Foot Rasps 脚锉

Nail Art Equipment 美甲设备

Nail Brushes 指甲刷

Nail Dryers 指甲烘干器

Nail Files 指甲锉

Nail Form 美甲指托

Nail Glue 指甲胶水

Sets & Kits 套装

Toe Separators 分趾器

Nail Treatments 指甲修复用品

Showing Shelf 美甲展示架

Templates 模板

Oral Hygiene 口腔清洁

Dental Flosser 牙线

Electric Toothbrush 电动牙刷

Interdental brush 牙间刷

Oral Irrigator 冲牙器

Teeth Whitening 牙齿美白产品

Toothbrush 牙刷

Toothbrush Sanitizer 牙刷杀菌器

Toothbrushes Head 牙刷头

Toothpaste 牙膏

Sanitary Paper 卫生用纸

Adult Diapers 成人尿布

Facial Tissue 面巾纸

Paper Napkins & Serviettes 餐巾纸

Toilet Tissue 卫生纸

Wet Wipes 湿巾

Sex Products 成人用品

Adult Games 成人游戏(骰子、扑克等)

Anal Sex Toys 后庭玩具

Catheters & Sounds 导管

Dildos 人造阴茎

Erotic Positioning Bandage 性爱绑带

Lasting Products 持久产品

Masturbators 男性自慰器

Medical Themed Toys 医疗玩具

Penis Rings 阴茎环

Pumps & Enlargers 阴茎增大

Safer Sex 安全/避孕

Condoms 避孕套

Female Contraceptives 女用避孕

Fertility Tests 排卵检测

Lubricants 润滑剂

Pregnancy Tests 早孕检测

Sex Dolls 性玩偶

Sex Furnitures 性交家具

Sex Toys 性玩具

Vibrators 震动器

Shaving & Hair Removal 剃须及脱毛产品

Aftershave 须后膏

Electric Shavers 电动剃须刀

Epilator 脱毛器

Hair Removal Cream 脱毛膏

Nose & Ear Trimmer 耳鼻毛器

Razor 剃须刀

Razor Blade 剃须刀片

Shaving Brush 剃须刷

Shaving Cream 剃须膏

Shaving Foam 剃须泡

Skin Care 护肤品

Body 身体护理

Creams 身体乳

Essential Oil 精油

Eyes 眼部护理

Creams 眼霜

Masks 眼膜

Face 脸部护理

Cleansers 洁面用品

Day Creams & Moisturizers 日霜

Emulsion 乳液

Facial Scrubs & Polishes 面部磨砂

Night Creams 晚霜

Toners 爽肤水

Treatments & Masks 面膜

Feet 足部皮肤护理

Hands & Nails 手部护理

Hand Creams & Lotions 护手霜

Hand Soaps 洗手液

Moisturizing Gloves 润肤手套

Nail Treatments 指甲护理

H Paraffin Baths 手蜡

Lips 唇部护理

Maternity 孕妇皮肤护理

Neck 颈霜

Sets 套装

Sun 防晒/助晒

After Sun Lotions 晒后修复

Body Self Tanners & Bronzers 身体美黑

Body Sunscreen 身体防晒

Facial Self Tanners & Bronzers 脸部美黑

Facial Sunscreen 脸部防晒

Lip Protection 唇部防晒

Skin Care Tool 护肤工具

Other Skin Care Tool 其他护肤工具

Paraffin Heater 蜡疗机

Wax Strip 蜡纸

Tattoo & Body Art 纹身及身体彩绘

Tattoo Grips 纹身手柄

Tattoo Guns 纹身机

Tattoo Inks 纹身色料

Tattoo Kits 纹身器具套件

Tattoo Needles 纹身针

Tattoo Power Supply 纹身电源

Tattoo Stencils 纹身模版

Tattoo Tips 纹身针嘴

Tattoo Accessories 纹身配件

Temporary Tattoos 一次性纹身

Tools & Accessories 工具/配件

Cotton Pads 化妆棉

Cotton Swabs 棉签

Foot Care Tool 足部护理工具

Hand Rests 手枕

Mirrors 镜子

Makeup Mirrors 化妆镜

Shower Mirrors 洗浴镜

Refillable Bottles & Accessories 分装瓶及
配件

Refillable Bottles 分装瓶

Toiletry Kits 工具套装

Computer& Office 电脑和办公

Computer Cables & Connectors
电脑连线及接插件

Computer Cleaners 电脑清洁用品

Computer Components 电脑组件和硬件

Add On Cards 扩展卡

CPUs 中央处理器

Computer Cases & Towers 台式机机箱

Fans & Cooling 散热系统

Floppy Drives 软驱

Graphics Cards 显卡

Internal Hard Drives 内置机械硬盘

Internal Memory Card Readers 内置读卡器

Internal Solid State Drives 内置固态硬盘

Motherboards 主板

Optical Drives 光驱

PC Power Supplies 电脑电源

RAMs 内存

Sound Cards 声卡

Video & TV Tuner Cards 采集卡和电视卡

Computer Peripherals 电脑外设

Computer Speakers 电脑音箱

Mice & Keyboards 鼠标键盘及配件

Digital Tablets 数位板

Keyboard Mouse Combos 键盘鼠标套装

Keyboards 键盘

Mouse 鼠标

Mouse Pads 鼠标垫 & 衬垫

Touch Pads 触摸板手写板

Monitors 显示器

CRT Monitors CRT 显示器

LCD Monitors 液晶显示器

Monitor Holder 显示器支架

Screen Protectors & Filters 屏幕保护膜和防窥膜

Touch Screen Panels 触摸屏面板

PC Game Hardware 电脑游戏外设

Flight Controls 飞行模拟控制

Gamepads PC 游戏手柄

Joysticks PC 游戏摇杆

Light Guns PC 游戏光线枪

Steering Wheels & Pedals PC 游戏方向盘

Printers 打印机

USB Gadgets USB 新奇特

USB Hubs USB Hub

Webcams 摄像头

Desktops 台式电脑

External Storage 移动硬盘,U 盘,刻录盘

Blank Disks 刻录盘

External Hard Drives 外置机械移动硬盘

External Solid State Drives 外置固态硬盘

HDD Enclosure 硬盘盒

Hard Drive Bags & Cases 硬盘壳包

USB Flash Drives U 盘

Industrial Computer & Accessories 工控产品

KVM Switches 切换器

KVM Switches 笔记本电脑附件

Keyboard Covers 键盘保护膜

LCD Hinges LCD 转轴

Lapdesks 笔记本桌和支架

Laptop Adapter 笔记本电源适配器

Laptop Bags & Cases 电脑包和壳

Laptop Batteries 笔记本电池

Laptop Cooling Pads 笔记本散热器

Laptop Docking Stations 插接站

Laptop LCD Inverter 笔记本 LCD 高压条

Laptop LCD Screen 笔记本液晶显示屏

Laptop Skins 笔记本美容贴纸

Laptop lock 笔记本锁

Replacement Keyboards 笔记本替换键盘

Screen Protectors 笔记本电脑屏保贴膜

Trackpoint Caps 指点帽

Laptops & Netbooks 笔记本与上网本

Laptops 笔记本电脑

Netbook 上网本

Mini PC mini 电脑

Networking 网络产品

3G Modems 3G 上网卡

3G/4G Routers 3G/4G 路由器

Access Points 无线热点

Firewall & VPN 防火墙和 VPN

Modems 调制解调器

Network Cabinets 网络机柜

Network Cards 网卡

Network Hubs 网络集线器

Network Switches 数据交换机

Networking Storage 网络存储

Networking Tools 网络用工具

Powerline Network Adapters
电力线网络适配器

Print Servers 打印服务器

Routers 路由器

USB Bluetooth Adapters/Dongles 蓝牙适配器/加密狗

WiFi Finder WiFi 信号探测器

Office Electronics 办公电子

3D Printers & 3D Scanners 3D 打印和 3D 扫描

3D Pens 3D 笔

3D Printer 3D 打印机

3D Printer Parts & Accessories 3D 打印零件
　及配件

3D Printing Materials 3D 打印耗材

3D Scanners 3D 扫描

All in One Printer 打印复印扫描一体机

Binding Machine 装订机

Calculators 计算器

Conference System 会议系统

Credentials Printer 证件打印

Electronic Dictionary 电子词典

Fax machines & Copiers 传真及复印

Copier 复印机

Digital Duplicator 速印机

Fax Machines 传真

Parts & Accessories 零件及配件

Laminator 塑封机

Laser Pens 激光笔

Network Print Servers 打印服务器

Paper Trimmer 裁纸器

Plotters 绘图仪

Printer Supplies 打印机耗材和部件

Cartridge Chip 墨盒/粉盒芯片

Continuous Ink Supply System 连续供墨系统

Fuser Film Sleeves 定影膜

Fuser Roller 定影辊

Ink Cartridges 墨水盒

Ink Refill Kits 填充墨水组

OPC Drum 光导鼓

Photo Paper 照相纸

Printer Control Cards 扩展卡

Printer Memory Modules 内存模块

Printer Parts 打印机部件

Printer Power Adapters 电源适配器

Printer Ribbons 打印机色带

Printer Trays 托盘

Toner Cartridges 墨粉盒、碳粉盒、粉仓

Toner Powder 墨粉、碳粉

Scanners 扫描设备

Shredder 纸机

Time Recording 考勤机

Visual Presenter 高拍仪

Servers 服务器

Software 软件

Tablets 平板电脑

Tablets & PDAs Accessories 平板电脑，
　PDA，MID 配件

Kids Tablet 儿童专用平板

Screen Protectors 平板\PDA 屏保贴膜

Screen Touch Gloves 触摸屏手套

Tablet Decals 平板电脑装饰贴纸

Tablet LCDs & Panels 平板电脑显示屏与
　面板

Tablet PC Stands 平板电脑支架

Tablets & e-Books Case 平板电脑和电子书
　保护套

Tablets Batteries & Backup Power 平板电脑
　电池和备用电源

Tablets Battery Chargers 平板电脑充电器

Tablets Pen 触摸笔

Workstations 工作站

Construction & Real Estate 建筑

Boards 板材

Aluminum Composite Panels 合成铝板，铝
　复合板，铝塑板

Cement Boards 水泥板

Plasterboards 石膏板

Sandwich Panels 夹芯板

Building Glass 建筑玻璃

Ceilings 吊顶，天花板

Ceiling Grid Components 天花龙骨零组件

Ceiling Tiles 吊顶天花板

Corner Guards 护角

Mouldings 线条

Countertops,Vanity Tops & Table Tops
　台面

Curtain Walls & Accessories 幕墙及附件

Curtain Wall Profiles 幕墙型材

Curtain Walls 幕墙

Spiders 驳接爪

Doors, Gates & Windows 门，窗

Door & Window Frames 门窗框

Door & Window Grates 门窗栅栏

Door & Window Screens 门窗纱

Door Sills 门槛

Doors 门

Gates 大门

Shutters 百叶门窗

Window Sills 窗沿

Windows 窗

Earthwork Products 土工材料

Geocells 土工格室

Geogrids 土工格栅

Geomembranes 土工膜

Geotextiles 土工布

Fireplaces, Stoves 壁炉，火炉

Fireplace Parts 壁炉零部件

Fireplaces 壁炉

Stoves 火炉

Floor Heating Systems & Parts 地暖系统及部件

Flooring & Accessories 地板及附件

Anti-Static Flooring 防静电地板

Bamboo Flooring 竹地板

Cork Flooring 软木地板

Engineered Flooring 复合地板

Flooring Accessories 地板附件

Plastic Flooring 塑料地板

Rubber Flooring 橡胶地板

Wood Flooring 实木地板

Functional Material 功能性材料

Fireproofing Materials 防火材料

Calcium Silicate Boards 硅酸钙板

Magnesium Oxide Boards 氧化镁板

Heat Insulation Materials 隔热材料

EPS Foam Boards 膨胀聚苯乙烯泡沫保温板

XPS Boards 挤塑式聚苯乙烯隔热保温板

Multifunctional Materials 多功能特殊材料

Soundproofing Materials 隔音材料

Acoustic Panels 吸音板

Sound Barriers 声屏障

Waterproofing Materials 防水材料

Roofing Felts 屋面毡

Waterproof Membrane 防水卷材

Garden Landscaping & Decking 园林景观与装饰

Other Landscaping & Decking 其他园林景观与装饰

HVAC Systems & Parts 冷暖通风系统及部件

Kitchen & Bath Fixtures 厨房和浴室设施

Bathroom Parts 卫浴设施零件

Bathtub Handrails 浴缸把手

Bathtub Pillows 浴缸枕头

Drains 去水，地漏

Filling Valves 进水阀

Flappers & Tank Balls 瓣阀、水箱球

Flush Valves 冲水阀

Plumbing Hoses 软管

Plumbing Nozzles 管口

Plumbing Traps 弯管

Tank Levers 水箱拉杆

Toilet Bowls 坐便器底座

Toilet Push Buttons 坐厕按钮

Toilet Seats 坐便器盖板

Bathroom Sinks 台盆

Bathroom Vanities 浴室柜

Bathtubs & Whirlpools 浴缸、按摩浴缸

Bidets 妇洗器

Kitchen Cabinets & Accessories 橱柜及附件

Kitchen Cabinets & Accessories Kitchen Cabinet Parts & Accessories 橱柜零配件

Kitchen Cabinets 橱柜

Kitchen Sinks 厨房水槽

Kitchen Storage 橱柜收纳件

Sanitary Ware Suite 卫浴套件

Shower Rooms & Accessories 淋浴房及附件

Bath Screens 浴屏/淋浴隔断

Shower Doors 淋浴门

Shower Rooms & Accessories Shower Rooms 淋浴房

Shower Trays 淋浴房底座

Solar Collectors 太阳能集热器

Spa Tubs & Sauna Rooms 水疗池与桑拿房

Sauna Rooms 桑拿房

Spa 水疗池

Squat Pans 蹲便器

Toilet Tanks 厕所水箱

Toilets 坐便器

Urinals 小便器

Ladders & Scaffoldings 梯子和脚手架

Scaffoldings 脚手架

Work Platforms 工作台

Landscaping Stone 环境石材

Cobbles & Pebbles 鹅卵石

Curbstones 路边石、镶边石、侧石、井栏石

Gravel & Crushed Stone 砾石，碎石

Mushroom Stone 蘑菇石

Paving Stone 铺路石

Tactile Paving 盲人石

Metal Building Materials 金属建材

Aluminum 铝材

Aluminum Bars 铝棒

Aluminum Pipes 铝管

Aluminum Powder 铝粉

Aluminum Profiles 铝型材

Aluminum Strips 铝带

Aluminum Wire 铝丝

Barbed Wire 带刺金属丝

Bearing Steel 轴承钢

Copper 铜材

Copper Bars 铜棒

Copper Cathode 电解铜

Copper Pipes 铜管

Copper Powder 铜粉

Copper Sheets 铜板

Copper Strips 铜带

Copper Wire 铜丝

Graphite Products 石墨制品

Graphite Crucible 石墨坩埚

Graphite Electrodes 石墨电极

Graphite Mold 石墨模具

Graphite Powder 石墨粉

Graphite Rod 石墨棒

Graphite Sheets 石墨板

Other Graphite Products 其它石墨产品

Ingots 金属锭

Aluminum Ingots 铝锭

Antimony Ingots 锑锭

Bismuth Ingots 铋锭

Copper Ingots 铜锭

Indium Ingots 铟锭

Lead Ingots 铅锭

Magnesium Ingots 镁锭

Manganese Ingots 锰锭

Other Ingots 其它锭

Silicon Ingots 硅锭

Steel Ingots 钢锭

Tin Ingots 锡锭

Titanium Ingots 钛锭

Zinc Ingots 锌锭

Iron Wire 铁丝

Magnetic Materials 磁性材料

Nickel 镍

Quartz Products 石英制品

Quartz Crucible 石英坩埚

Quartz Plate 石英板

Quartz Rods 石英棒

Quartz Stone 石英石

Quartz Tubes 石英管

Rare Earth & Products 稀土及其制品

Spring Steel 弹簧钢

Stainless Steel 不锈钢材

Stainless Steel Angles 不锈钢角钢

Stainless Steel Balls 不锈钢球

Stainless Steel Bars 不锈钢棒

Stainless Steel Channels 不锈钢槽钢

Stainless Steel Flats 不锈钢扁钢

Stainless Steel Pipes 不锈钢管

Stainless Steel Sheets 不锈钢板

Stainless Steel Strips 不锈钢带

Stainless Steel Wire 不锈钢丝

Steel 钢

Steel Channels 槽钢

Steel Pipes 钢管

Steel Rails 钢轨

Steel Rebars 螺纹钢

Steel Round Bars 圆钢

Steel Sheets 钢板

Steel Strips 带钢

Steel Structures 钢结构

Steel Wire 钢丝

Structural Steel 结构钢

Tinplate 马口铁

Titanium 钛材

Titanium Bars 钛棒

Titanium Foil 钛箔

Titanium Pipes 钛管

Titanium Powder 钛粉

Titanium Sheets 钛板

Titanium Wire 钛丝

Tungsten 钨

Tungsten Bar 钨棒

Tungsten Crucibles 钨坩埚

Tungsten Foil 钨膜

Tungsten Pipes 钨管

Tungsten Powder 钨粉

Tungsten Sheets 钨板

Tungsten Wire 钨丝

Wire Mesh 金属丝网

Aluminum Wire Mesh 铝丝网

Copper Wire Mesh 铜丝网

Iron Wire Mesh 铁丝网

Other Wire Mesh 其他丝网

Steel Wire Mesh 钢丝网

Other Home Improvement 其他家装

Other Home Improvement 塑料建筑材料

Architecture Membrane 建筑膜

Plastic Profiles 塑料型材

Sun Sheets & PC Embossed Sheets 阳光板，
PC 颗粒板

Quarry Stone & Slabs 荒料及石板材

Artificial Stone 人造石

Natural Stone 天然石材

Basalt 玄武岩

Granite 花岗岩

Limestone 石灰石

Marble 大理石

Sandstone 砂岩

Slate 板岩

Stairs & Stair Parts 楼梯及部件

Balustrades & Handrails 栏杆，扶手

Stair Parts 楼梯零部件

Stairs 楼梯

Stone Carvings and Sculptures 石雕石刻

Arches 拱门

Pillars 柱子

Statues 雕像

Stone Garden Products 石制园林产品

Stone Reliefs 浮雕

Sunrooms & Glass Houses 阳光房，玻璃房

Tiles & Accessories 瓷砖及附件

Mosaics 马赛克

Tile Accessories 瓷砖附件

Tiles 瓷砖

Timber 木材

Anticorrosive Woods 防腐木

Decorative High-Pressure Laminates / HPL
防火板

Fibreboards 纤维板/密度板

Finger Jointed Boards 指接板/集成材

Flakeboards　刨花板

Laminated Wood Boards / Blockboards　细木
　工板/大芯板

Melamine Boards　三聚氰胺板

Plywoods 胶合板

Solid Wood Boards　实木板

Veneers　木皮

Consumer Electronics　消费电子

Accessories & Parts　零配件

3D Glasses 3D 眼镜

Battery Storage Boxes　电池收纳盒

Blank Records & Tapes　空白磁带和卡带

CD Bags & Cases CD 包、CD 盒

CD/DVD　播放器箱包

Cable Winder　绕线器

Camera Strap　相机背带

Camera/Video Bags　照相机,摄像机包/袋

Card Readers　读卡器

Chargers　充电器

Digital Batteries　电池

Digital Cables　数码线材

Audio & Video Cables　音频/视频线

Data Cables　数据线

Power Cables　数码产品充电线

Earphone Accessories　耳机配件

Earphones & Headphones　耳机

Electrical Plugs & Sockets　插座、插头

MP3/MP4 Bags & Cases MP3/MP4 播放器
　保护套

Memory Card Cases　内存卡盒

Memory Cards　储存卡

Microphones　麦克风

Projector Bulbs　投影仪灯泡

Remote Control　遥控器

Screen Cleaners　屏幕清洁器

Screen Protectors　屏幕保护膜

Speakers　扬声器

TV Mounts　电视机支架

Video Game Player Cases　游戏机包,保护套

Video Glasses　视频眼镜

Camera & Photo 摄影摄像

Backgrounds　摄影背景

Battery Grip　相机电池手柄

Camcorder Lens　摄像镜头

Camera Cleaning　相机清洁用品

Camera Drones & Accessories　航拍飞行器
　和配件

Camera Drones & Accessories Camera Drone
　Accessories　航拍飞行器配件

Camera Drones　航拍飞行器

Camera Filters　相机滤镜

Camera Lens Hoods　镜头遮光罩

Camera Lenses　相机镜头

Digital Cameras　数码相机

Digital Microscope　数码显微镜

Digital Photo Banks　数码伴侣

Digital Photo Frames　数码相框

Film Cameras　胶片相机

Films　胶卷

Flash Diffuser　柔光罩

Flashes　闪光灯

Instant Camera　即拍即得相机

Len Caps　镜头盖

Lens Adapter　镜头转接环

Mini Camcorders　微型摄像机

Photo Studio Accessories　摄影工作室用品

Photographic Lighting　摄影灯

Shutter Release　快门线和遥控器

Sports & Action Video Cameras　运动摄像机

Sports Camcorder Accessories　运动摄像机
　配件

Tripod & Accessories　三脚架和配件

Selfie Sticks　自拍杆

Tripod Heads　云台

Tripod Legs　三脚架腿

Tripod Monopods　快装板

Tripods　三脚架

Video Cameras　摄像机

DIY Parts DIY 电子产品

Integrated Circuits 集成电路

LCD Displays LCD 显示屏

LED Displays LED 显示屏

Sensors 传感器

Electronic Cigarettes 电子烟

Electronic Cigarette Accessories 电子烟配件

Electronic Cigarette Kits 电子烟套装

Electronic Cigarette Parts 电子烟零件

Electronic Cigarette Atomizer Core 雾化器芯

Electronic Cigarette Atomizers 雾化器

Electronic Cigarette Battery 电源&供电模块

Electronic Cigarette Chargers 充电器

Electronic Cigarette Mods 电池仓

Gaming & Accessories 游戏机及其附件

Arcade Sticks 游戏摇杆

Controllers 游戏手柄

DC/MD/SS/SEGA 配件

Cables AV 线

Controllers 手柄

Memory Cards 记忆卡

Power Supply 电源

Dance Pads 跳舞毯

Gaming Accessories 游戏配件附件

Accessories For NDS   NDS 配件

Accessories For PS2   PS2 配件

Accessories For PS3   PS3 配件

Accessories For PSP   psp 配件

Accessories For Wii   wii 配件

Accessories For XBOX   XBOX 配件

Handheld Game Players 掌上游戏机

NDSI/NDSL   NDS 配件

Other Accessories 其他配件

Power Supply 电源

Stylus 触笔

NGC/FC/N64/SFC 配件

Cables AV 线

Controllers 手柄

Memory Cards 记忆卡

Power Supply 电源

Nintendo 配件

Buttons 按键

Cases 外壳

Screens 屏幕

Bags & Soft Cases 专用布袋/软套/硅胶套

Cables 各类线材

Earphones 专用耳机

Hard Cases 水晶盒/铁壳

PSV   PSV 配件

Batteries   PSV 电池

Cases PSV 保护套/外壳

Chargers PSV 充电器

Earphones PSV 专用耳机

Screen Protectors PSV 屏幕贴膜

Stickers PSV 彩贴/贴纸

PlayStation   PS 配件

Cables 各类线材

Chargers 座充

Console Bags 主机包

Controllers 手柄

Fans 风扇

Guitars 吉他

Power Supply 电源

Stands 支架

Video Game Consoles 视频游戏机

Wheels 游戏方向盘

Wii   Wii 配件

Arcade Sticks 格斗摇杆

Balance Board FIT 平衡板

Boxing Gloves 拳击手套

Dance Mats 跳舞毯

Drums 鼓

Flight Controller 飞行摇杆

Golf Stick 高尔夫球杆

Guitars 吉他

Guns 游戏枪

Hockey Sticks 曲棍球杆

Memory Cards WII 专用记忆卡

Motion Plus　加速器

Motorboat　摩托艇

Musical Packs　音乐套装

Nun chucks　右手手柄

Ping Pong Paddles　乒乓球拍

Remotes　左手手柄

Snooker Cue　桌球杆

Sports Packs　运动套装

Swords　光剑

Yuga Mats　瑜伽垫/健身毯

XBOX　XBOX 配件

Batteries & Chargers　电源

Harddisk Cases　硬盘盒

Kinect　体感器

Home Audio & Video Equipments 家用音视频设备

Amplifiers　功率放大器,扩音机

Blu-ray Players　蓝光播放机

Cassette Recorders & Players　磁带录音播放机

DVD, VCD Players　DVD, VCD 播放机

Digital Voice Recorders　数码录音笔

HDD Players　硬盘播放机

Home Theatre System　家庭影院系统

Karaoke Players　卡拉 OK 播放机

Projects & Accessories　投影及配件

Projection Screens　投影屏

Projector Brackets　投影支架

Projectors　投影仪

Radio　收音机

TV Stick　电视棒

Televisions　电视机

Tv Receivers　电视接收

Radio & TV Broadcasting Equipment 无线广播&电视广播设备

Satellite TV Receiver　卫星电视接收器

Set Top Box　机顶盒

TV Antenna　电视天线

Portable Audio & Video　便携播放器,阅读器

CD Players CD　播放机

Headphone Amplifier　耳机功率放大器

MP3 Players MP3　播放器

MP4 Players MP4　播放器

Quran Players　古兰经播放器

e-Book Readers　电子书阅读器

Portable HiFi　便携 HiFi

HiFi Accessories HiFi 配件

HiFi Amplifiers HiFi 功放

HiFi Earphones & Headphones HiFi 耳机

HiFi Players HiFi 播放器

Power Source　电池、电源

Adapers　电源适配器

Battery Packs　电池组

Converters　变频器

Rechargeable Batteries　充电电池

Solar Panel　太阳能电池

Switching Power Supply　开关电源

Smart Electronics　智能电子

Smart Health　智能健康

Smart Fitness　智能健身设备

Smart Scales　智能秤/体重管理设备

Smart Home　智能家庭

Home Automation Kits 家居自动化套件/总成

Home Automation Modules 家居自动化模块

Smart Access Lock　智能锁

Smart CCTV　智能监控

Smart Finder　智能寻物

Smart Home Controls　智能开关和智能遥控

Smart Home Illumination　智能照明

Smart Home Sensor　智能家庭传感器

Smart Power Socket Plug　智能电源插座

Smart Thermometer　智能温度计

Smart Remote Control　智能遥控产品

Wearable Devices　智能穿戴设备

Accessories　智能配饰

Activity Trackers　运动跟踪

Glasses　眼镜

Smart Watches　手表

Wristbands　手环

Customized Products 定制化产品

Apparel & Fashion Accessories 服饰

Aprons 围裙

Baby Clothes 婴幼儿服饰

Bed and Bath Apparel 卫浴寝具服装

Costumes 扮演类服装

Customize Sweaters 可定制毛衣

Customize Wedding Dresses 定制婚纱

Fashion Accessories 饰品

Hoodies & Sweatshirts 卫衣/帽衫

Other Apparel 其他服装

Shirts 衬衫

Sportswear 运动服

T-Shirts T 恤

Compressed T-Shirts 压缩 T 恤

EL T-Shirts ELT 恤

Polo Shirts Polo 衫

Round Neck T-Shirts 圆领 T 恤

V-neck T-Shirts V 领 T 恤

Tank Top 背心

Uniforms 制服

Vests & Waistcoats 马夹 / 外套背心

Work Wear 工作服

Bags & Hats 包和帽子

Hats & Caps 帽子

Baseball Caps 促销品棒球帽

Bucket Hats 促销品渔夫帽

Knitted Hats 针织帽

Other Hats & Caps 其它促销品帽子

Party Hats 促销品节日帽

Straw Hats 促销品草帽

Sun Visors 促销品遮阳帽

Baseball Sport Cap 棒球运动帽

Bracelets & Cuff 手镯手链

BBB 手镯迁移

Electronic Products 电子产品

Car Electronics 车载电子

Card Reader 读卡器

Cell Phone Holders 手机座

Earphone / Headphone 耳机

Keyboard 键盘

MP3 Player MP3 播放器

Memory Card 存储卡

Mouse 鼠标

Phone Cases 手机壳

Promotional Mouse Mat 促销鼠标垫

Speakers 扩音器

Tablet Cases 平板电脑壳

USB Flash Drive U 盘

USB Gadget USB 小配件

Faucet 龙头

Hair Braids 发辫

Hard Drives 硬盘

Household Products 促销家居用品

Badge 促销徽章

Barware 促销用酒吧用具

Coaster 促销杯垫

Cooking Tool 促销炊具

Drinkware 促销用饮具

Mugs 促销马克杯

Other Promotional Drinkware 其他促销饮具

Sport Bottles 促销运动水壶

Travel Mugs 促销旅行杯

Vacuum Flasks 促销保温杯

Fans 促销扇子

Food & Beverage 促销用食品饮料

Glassware 玻璃器皿

Home Storage & Organization 促销家用
收纳

Key Chains 促销钥匙扣

Magnet 促销磁贴

Mirror 促销用镜子

Mobile Phone Chain 促销手机链

Opener 促销开瓶罐器

Pet Products 促销宠物用品

Photo Frame 促销相框

Tableware 餐具

Dinnerware 促销用餐具

Other Promotional Tableware 其它促销餐具

Table Decoration & Accessories 促销用餐桌饰品与附件

Tape Measures 促销卷尺

Towel 促销品毛巾

Mounts 支架

Novelty Products 新奇特产品

Umbrella 促销雨伞

Outdoor Sports 户外运动

Care 护肤

Calendar 促销日历

Cards 促销卡片

Indoor Sports 室内运动

Paper Packing 促销纸包装

Printing Products 宣传印刷品

Personal Care 促销用个人护理产品

Bath Products 促销用沐浴产品

Brushes 促销用化妆刷

Combs 促销梳子

Cosmetics 促销用化妆品

First Aid Kits 促销用急救包

Hand Sanitizers 促销用洗手液

Lip Balm 促销用润唇膏

Nail Clippers 促销用指甲钳

Promotional Pen 促销笔

Ballpoint Pen 促销圆珠笔

Fountain Pen 促销钢笔

Gel Pen 促销中性笔

Laser Pointer 促销激光笔

Marker Pen 促销马克笔

Radio Pen 促销无线电笔

Promotional Stationery 促销文具

Calculator 促销计算器

Desk Organizer 促销桌面收纳

Lanyard 促销挂绳

Memo Pad 促销便签本

Notebook 促销笔记本

Promotional Toys & Crafts 促销玩具工艺品

Event & Party Supplies 促销庆典派对用品

Souvenirs & Awards 促销纪念品

Watch 促销手表

Promotional Toys & Crafts 摩托车防护用具

School Supplies 学习用品

Correction Fluid 修正液

School Supplies 母婴学习用品

Sewing Supplies 缝纫用品

Threads 纺线

Sports Accessories 专业运动配饰

Wearable Electronic Devices 可穿戴式电子设备

Cameras 相机

Swimming 游泳

Customized Swimming 定制游泳

Team Sports 团体运动

Mens Leggings 男打底裤

Textiles & Leather 纺织及皮革(原材料)

Textile Accessories 纺织辅料

Textile Accessories Textile 纺织辅料叶子

Warning Flags 警示彩旗

Wigs 头套

Winter Sports 冬季运动

Bags 包

Backpacks 促销品背包

Bag Accessories 促销品包配件

Cosmetic Bags 促销品化妆包

Handbags 促销定制手提包

Laptop Bags 促销品电脑包

Laptop Sleeves 促销品电脑内胆包

Non-woven Bags 促销品无纺布袋

Other Tote Bags 其它促销品购物袋

Paper Bags 促销品纸袋

Plastic Bags 促销品塑料购物袋

Wallets, Change purses & Accessories 钱包、零钱包和附件

Electrical Equipment & Supplies 电气设备和产品

Batteries 电池

Battery Packs 电池组

Button Cell Batteries 纽扣电池

Fuel Cells 燃料电池

Other Batteries 其他电池

Primary & Dry Batteries 一次性电池

Rechargeable Batteries 充电电池

Solar Cells, Solar Panel 太阳能电池

Solar Energy Systems 太阳能系统

Storage Batteries 蓄电池

Circuit Breakers 电路断路器

Connectors & Terminals 连接器和接线端

Alligator Clips 鳄鱼夹

Connectors 连接器

Solar Controllers 太阳能控制器

Terminal Blocks 线弧，接头排

Terminals 接线端子

Contactors 接触器

Electronic & Instrument Enclosures 机柜

Fuse Components 熔断器/保险丝元件

Fuses 保险丝，熔断器

Generators 发电机

Alternative Energy Generators 风力，水力，太阳能发电机

Diesel Generators 柴油发电机

Gas Turbine Generators 燃气发电机

Gasoline Generators 汽油发电机

Power Distribution Equipment 配电输电设备

AC/DC Adapters 电源适配器

Inductors 电感器

Industrial Power Supply 工业专用电源

Inverters & Converters 变极器和变频器

Other Power Supplies 其它电源

Switching Power Supply 开关电源

Uninterrupted Power Supply (UPS) 不间断电源

Voltage Regulators/Stabilizers 稳压电源

Professional Audio, Video & Lighting 专业音响、视频、灯光

Relays 继电器

DIP Switches 拨码开关

Flow Switches 流量开关

Limit Switches 限位开关

Pressure Switches 压力开关

Push Button Switches 按钮开关

Remote Control Switches 遥控开关

Rocker Switches 摇臂开关

Rotary Switches 旋转开关

Slide Switches 滑动开关

Time Switches 时间开关

Toggle Switches 扳手开关

Wall Switches 墙壁开关

Transformers 变压器

Wires, Cables & Cable Assemblies 电线，电缆，电缆组件

Cable Manufacturing Equipment 电线电缆制造设备

Control Cables 控制电缆

Electrical Wires 电线

Instrumentation Cables 仪器仪表电缆

Power Cables 电力电缆

Power Cords & Extension Cords 电源线，延长线

Wiring Harness 线束

Cable Clips 电缆夹

Cable End Caps 电缆端盖

Cable Glands 电缆固定头

Cable Sleeves 电缆套管

Cable Ties 电缆扎带

Cable Trays 电缆桥架

Tie Mounts 扎线固定座

Wiring Ducts 线槽

Electronic Components & Supplies 电子元器件

Electronic Components & Supplies 电子器件/有源元件

Diodes 晶体二极管

Filters 滤波器

Integrated Circuits 集成电路

Oscillators 振荡器

Rectifiers 整流器

Resonators 共振器，谐振器

Sensors 传感器

Thyristors 半导体闸流管

Transistors 晶体管

EL Products 电子发光产品

Electronic Accessories & Supplies 电子材料和附件

Electrical Ceramics 电工陶瓷

Electrical Contacts and Contact Materials 电触点及材料

Insulation Materials & Elements 绝缘材料

Keypads & Keyboards 按键和键盘

Semiconductors 半导体

Electronic Data Systems 电子数据系统

Electronic Signs 电子标识

Electronics Production Machinery 电子元器件制造设备

Electronics Stocks 电子库存

Optoelectronic Displays 显示器件

LCD Modules LCD 液晶模块

LED Displays LED 显示屏

Passive Components 电子元件/无源元件

Acoustic Components 电声器件

Capacitors 电容器

Dining Room Furniture 餐厅家具

Dining Chairs 餐椅

Dining Room Sets 餐厅套装

Dining Tables 餐桌

Sideboards 餐具柜

Kitchen Furniture 厨房家具

Living Room Furniture 客厅家具

Bookcases 书柜，书架

CD Racks CD 架

Chaise Lounge 躺椅

Coat Racks 衣帽架

Coffee Tables 茶几，小边桌

Console Tables 角桌

Living Room Cabinets 客厅柜子

Living Room Chairs 客厅椅

Living Room Sets 客厅套装

Living Room Sofas 客厅沙发

Shoe Racks 鞋架，鞋柜

Stools & Ottomans 凳子，脚凳

TV Stands 电视柜

Beach Chairs 沙滩椅，户外椅

Garden Chairs 花园椅

Garden Sets 花园套装，户外套装

Garden Sofas 花园沙发，户外沙发

Hammocks 吊床

Other Outdoor Furniture 其他户外家具

Outdoor Tables 户外桌，花园桌，野餐桌

Patio Benches 户外长椅

Patio Swings 户外秋千

Patio Umbrellas & Bases 户外伞，基座

Sun Loungers 户外躺椅

Adhesives 黏合剂

Caulk 缝隙胶

Epoxies 环氧树脂胶

Sealers 涂封物

Silicone Sealant 硅密封胶

Tape 胶带

Tile Grout 瓷砖填缝剂

Wood Glue 木胶

Brackets 托架

Chains 链条

Clamps 夹具

Door Hardware 门五金及锁

Automatic Door Operators 门自动启闭装置

Door Bolts 门插销

Door Closers 闭门器

Door Handles 门拉手

Door Hinges 门合页/铰链

Door Plates 门牌

Door Rollers 门辊轴

Door Stops 门掣/门顶/门阻器/门吸

Door Viewers 猫眼

Doorbells 门铃

Doorknobs 圆筒式门把

Glass Clamps 玻璃夹

Elevators & Elevator Parts 电梯及部件

Elevator Parts 电梯零部件

Elevators 电梯

Escalators & Escalator Parts 自动扶梯及零部件

Escalator Parts 自动扶梯零部件

Escalators 自动扶梯

Moving Walks 自动人行道

PCB & PCBA 印制(刷)电路板和印制(刷)电路板组件

Double-Sided PCB 双面板

FPC 柔性电路板 / 挠性印刷电路板

Multilayer PCB 多层板

Rigid PCB 硬性电路板 / 普通印刷电路板

Single-Sided PCB 单面板

Potentiometers 电位器

Resistors 电阻器

Food 食品

Coffee 咖啡

Dried Fruit 干果

Dried Goods/Local Specialties 干货/土特产

Grain Products 谷物制品

Medlar 枸杞

Nut & Kernel 坚果

Tea 茶叶

Black Tea 红茶

Fruit Tea 水果茶

Grain Tea 谷物茶

Green Tea 绿茶

Herbal Tea 草本茶

Oolong Tea 乌龙茶

Other Tea 其他茶

Pu'Er Tea 普洱茶

White Tea 白茶

Yellow Tea 黄茶

flower tea 花茶

Children Furniture 儿童家具

Children Beds 儿童床

Children Cabinets 儿童柜

Children Chairs 儿童椅

Children Furniture Sets 儿童家具套装

Children Tables 儿童桌

Commercial Furniture 商用家具

Bar Furniture 酒吧家具

Bar Chairs 吧椅

Bar Furniture Sets 酒吧家具套装

Bar Stools 酒吧凳

Bar Tables 酒吧桌

Hotel Furniture 酒店家具

Hotel Bedroom Sets 酒店卧室家具套装

Hotel Beds 酒店床

Hotel Chairs 酒店椅

Hotel Sofas 酒店沙发

Hotel Trolley 酒店推车

Luggage Racks 行李架

Laboratory Furniture 实验室家具

Library Furniture 图书馆家具

Restaurant Furniture 饭店家具，餐馆家具

Restaurant Chairs 饭店椅

Restaurant Sets 饭店家具套装

Restaurant Tables 饭店桌

Salon Furniture 沙龙家具

Barber Chairs 理发椅

Massage Tables & Beds 按摩桌，按摩床

Nail Tables 指甲桌

Salon Trolley 沙龙推车

Shampoo Chairs 洗发椅

School Furniture 学校家具

Dormitory Beds 宿舍床

Other School Furniture 其它学校家具

School Chairs 学校椅子

School Desks 学校桌

School Sets 学校桌椅套装

Theater Furniture 影剧院礼堂家具

Waiting Chairs 等待座椅

Chair Mechanisms (坐具类)机械伸展装置

Furniture Frames 家具构架

Furniture Legs　家具腿脚

Swivel Plates　转盘

Home Furniture　家用，民用家具

Bathroom Furniture　浴室家具

Bedroom Furniture　卧室家具

Bedroom Sets　卧室套装

Beds　床

Dressers　梳妆台

Mattresses　床垫

Nightstands　床头柜

Wardrobes　衣柜,衣橱

Knockers　门环

Fasteners　紧固件

Anchors　锚栓件

Bolts　螺栓

Dowel　定位销

Grommets　扣眼

Nails　钉

Nut & Bolt Sets　螺母螺栓组合

Nuts　螺母

Picture Hangers　照片墙钉

Pins　销子

Plugs　止付螺钉

Rivets　铆钉

Screws　螺丝钉

Staples U 型钉

Tacks　大头钉

Threaded insert　螺纹嵌件

Washers　垫圈

Furniture Hardware　家具五金

Cabinet Catches　柜门吸

Corner Brackets　角码

Furniture Bolts　家具插销

Furniture Casters　家具脚轮

Furniture Hinges　家具合页/铰链

Furniture Locks　家具锁

Handles & Knobs　家具把手

Slides　滑轨

Garage Door Hardware　车库门五金

Hasps　搭扣

Hooks　吊钩

Locks　锁具

Other Door Hardware　大门五金

Other Hardware　其他五金件

Springs　弹簧

Windows Hardware　窗户五金

Latches　窗户插销

Sealing Strips　密封条

Wind Brace　风撑

Window Rollers　窗户滑轮

Window-Dressing Hardware　窗饰五金

Home & Garden　家居用品

Arts,Crafts & Sewing　手工艺品&缝纫用品
　　(半成品)

Apparel Sewing & Fabric　服装缝纫用品及
　　面料

Badges　徽章

Buckles　扣环

Buttons　纽扣

Cords　绳子

Down　羽绒

Embroidery Machines　刺绣机械

Fabric　面料

Feather　羽毛

Fiber　纤维

Fur　皮毛

Garment Beads　服装用散珠

Garment Clips　服装用夹子

Garment Eyelets　服装用气眼

Garment Hooks　裙裤钩

Garment Labels　服装用标签

Garment Rivets　服装用铆钉

Garment Tags　服装用挂牌

Genuine Leather　真皮

Interlinings & Linings　夹层及衬里

Lace　花边

Mannequins　服装人体模特儿

Overlockers　包缝机

Patches 服装用贴补片

Rhinestones 烫钻

Ribbons 缎带 / 丝带

Sequins 散装亮片

Sewing Machines 缝纫机

Sewing Needles 缝纫针

Sewing Threads 缝纫线

Shoulder Pads 肩垫

Stopper 吊钟/卡扣

Synthetic Leather 合成皮革

Tag Guns 标签枪

Tailor's Scissors 裁缝剪

Tassel Fringe 流苏

Thread 线

Trimming 排须

Velcro Tapes 粘扣带

Webbing 织带

Yarn 纱

Zipper Sliders 拉练头

Zippers 拉链

Needle Arts & Craft 家居手工缝纫用品

Cross-Stitch 十字绣

Diamond Painting Cross Stitch 贴钻绣

Embroidery 刺绣

Sewing Tools & Accessory 缝纫工具及附件

Needle Arts & Crafts 家居缝纫用品

Painting Supplies 绘画用品

Scrapbooking & Stamping 剪贴簿及印章

Stamps 印章

Bathroom Products 浴室用品

Basins 脸盆

Bath Brushes, Sponges & Scrubbers
  浴球，浴刷，浴棉

Bath Mats 浴室防滑垫

Bath Pillows 浴枕

Bathroom Sets 浴室套装

Household Scales 家用秤

Shower Caps 浴帽

Shower Curtain Poles 浴帘杆

Shower Curtains 浴帘

Toilet Plungers 马桶吸盘

Toilet Seat Cover 马桶盖套

Tubs 浴盆

Collectibles 收藏品

Advertising 广告相关

Coins 钱币

Commemorative Coins 纪念币

Genuine Coins 收藏类钱币(真币)

Reproduction Coins 仿真收藏钱币

Militaria 军备

Badges 徽章

Field Equipment 军用装备

Medals and Awards 奖牌和奖状

Swords and Blades 剑和刀具

Uniforms 制服

Postage Stamps 邮票(真正发行的邮票)

Festive & Party Supplies 节庆派对用品

Christmas Decoration Supplies 圣诞装饰

Decorative Flowers & Wreaths 装饰花, 花环

Event & Party Supplies 庆典派对用品

Other Holiday Supplies 其它节日用品

Party Masks 派对面具

Fertilizer 肥料

Biological Fertilizer 生物肥

Compound Fertilizer 复合肥

Nitrogen Fertilizer 氮肥

Organic Fertilizer 有机肥

Phosphate Fertilizer 磷肥

Plant Food 观赏花卉用肥

Potassium Fertilizer 钾肥

Garden Buildings 园林建筑及装饰

Arches, Arbours, Pergolas & Bridge 拱门,棚
  架,凉亭与桥

Cabins & Garden Rooms 休闲屋

Fencing, Trellis & Gates 栅栏,格架和门

Garages, Canopies & Carports 车棚

Garden Greenhouses 花园暖房

Garden Ornaments 花园装饰品

Playhouses　儿童游戏房

Sheds & Storage　工具房和储物间

Summerhouses　度假屋

Garden Pots & Planters　花盆与种植

Basket Liners　吊篮垫

Bonsai　盆栽

Flower Pots & Planters　花盆

Grow Bags　成长袋

Hanging Baskets　吊篮

Nursery Pots　育苗盆

Nursery Trays & Lids　育苗托盘和盖子

Plant Seeds　种子

Pot Trays 花盆托盘

Garden Tools　园林工具

Chainsaws　链锯(油锯)

Cleaning Tools　清洁工具

Combination　园林工具组合套装

Digging Tools　园艺挖掘工具

Fork　叉

Garden Blower & Vacuum　园林鼓风机和吸尘器

Garden Cultivator　中耕工具

Garden Shredder　园林粉碎机

Garden Gloves　手套

Grass Trimmer　割灌机(打草机)

Hedge Trimmer　绿篱机

Hoe　锄头

Lawn Mower　割草机

Pickaxe　镐

Pruning Tools　修剪工具

Rake　耙

Sickle　镰刀

Spade & Shovel　锹、铲

Trowel　镘

Greenhouses　温室

Household Thermometers　家用温度计

Mailboxes　邮箱

Outdoor Heaters　户外取暖用品

Chimeneas　户外暖炉

Fire Pits　火盆

Patio Heaters　庭院取暖器

Pest Control　虫害防治

Shade　遮阳用品

Awnings　遮阳篷

Gazebos　遮阳棚

Shade Accessories　遮阳用品附件

Shade Sails & Nets　遮阳布

Watering & Irrigation　花园灌溉用品

Garden Hoses & Reels　园林水管和水管车

Garden Sprinklers　园林洒水装置

Garden Water Connectors　园林水连接件

Garden Water Guns　园林水枪

Garden Water Timers　园林灌溉计时器

Sprayers　喷雾器

Water Cans　洒水壶

Watering Kits　灌溉套装

Blinds, Shades & Shutters　百叶帘和帘子

Candle Holders　烛台

Candles　蜡烛

Clocks　钟

Alarm Clocks　闹钟

Antique Style Clocks　仿古工艺钟

Clock Parts & Accessories　钟的零部件和附件

Desk & Table Clocks　座钟，台钟

Digital & Analog-Digital Clocks　数字和模拟钟

Floor Clocks　落地钟

Hourglasses　沙漏

Mechanical Clocks　机械钟

Specialty Clocks　特殊用钟

Wall Clocks　壁挂钟

Crystal Soil　水晶泥

Curtain Poles, Tracks & Accessories　窗杆,窗轨与附件

Decoration Crafts　家居装饰工艺品

Bamboo Crafts　竹工艺品

Clay Crafts　黏土工艺品

Crystal Crafts　水晶工艺品

Glass Crafts 玻璃工艺品

Lacquerware 漆器工艺品

Leather Crafts 皮革工艺品

Metal Crafts 金属工艺品

Natural Crafts 天然工艺品

Paper Crafts 纸工艺品

Plastic Crafts 塑料工艺品

Pottery & Enamel 陶瓷,珐琅

Resin Crafts 树脂工艺品

Semi-Precious Stone Crafts 宝石工艺品

Stone Crafts 石头工艺品

Textile & Fabric Crafts 纺织工艺品

Wicker Crafts 编结工艺品

Wood Crafts 木制工艺品

Decorative Films 装饰贴膜

Fireplace Sets & Accessories 壁炉用具

Flags, Banners & Accessories 旗帜,横幅

Frame 画框,相框,镜框等

Fridge Magnets 冰箱贴

Incense 熏香

Incense Burners 熏香炉

Mirrors 镜子

Money Boxes 储蓄罐

Music Boxes 音乐盒

Painting & Calligraphy 书法绘画

Photo Albums 相册

Plaques 饰板

Screens & Room Dividers 屏风和隔断

Vases 花瓶

Wall Sticker 墙贴

Window Films 窗户贴膜

Bag Clips 袋夹

Hooks & Rails 钩子和栏杆

Saran Wrap & Plastic Bags 保鲜膜 &保鲜袋

Shoe Hanger 鞋架

Storage Bags 收纳包

Storage Baskets 收纳篮

Storage Bottles & Jars 收纳瓶和收纳罐

Storage Boxes & Bins 收纳盒和收纳箱

Storage Drawers 收纳抽屉

Storage Holders & Racks 收纳架

Storage Trays 收纳托盘

Home Textile 家纺成品

Bed Skirt 床裙

Bedding Set 床上成套纺织用品

Bedspread 床罩

Blanket 毯子

Carpet 地毯

Chair Cover 椅套

Comforter 胖被子

Curtain 窗帘

Cushion 坐垫/靠垫

Cushion Cover 垫套

Duvet Cover 被套

Handkerchief 手帕

Mat 地垫

Mattress Cover 床垫套

Mosquito Net 蚊帐

Other Home Textile 其他家纺用品

Pillow 枕头

Pillow Case 枕套

Quilt 缝被

Rug 小地毯

Sashes 椅子飘带

Sheets 床单

Sofa Cover 沙发套

Table Cloth 桌布

Table Napkin 餐巾

Table Runner 桌旗

Table Skirt 桌裙

Tapestry 挂毯

Throw 盖毯

Towel 毛巾/浴巾

Valance 短帷幕

Bathroom Accessories 卫浴五金件

Bath Hardware Sets 卫浴五金套件

Bath Mirrors 浴室镜

Bathroom Shelves 置物架

Cup & Tumbler Holders 杯架

Grab Bars 扶杆

Liquid Soap Dispensers 皂液器

Paper Holders 纸巾架

Robe Hooks 衣挂钩

Soap Dishes 皂碟

Toilet Brush Holders 马桶刷

Towel Bars 毛巾杆

Towel Racks 毛巾架

Towel Rings 毛巾环

Faucets, Mixers & Taps 龙头，阀门

Aerators 起泡器

Basin Faucets 台盆龙头

Bath & Shower Faucets 浴缸淋浴龙头

Bib cocks 水嘴

Bidet Faucets 妇洗器龙头

Faucet Cartridges 龙头阀芯

Kitchen Faucets 厨房龙头

Shower Head 花洒头

Wallpapers 墙纸

Aprons 围裙

Brooms & Dustpans 扫帚和畚箕

Buckets 桶

Cleaning Brushes 清洁刷

Cleaning Cloths 清洁布

Dusters 掸子

Household Chemicals 家用日化产品

Air Fresheners 空气清新剂

Detergent 洗涤剂

Drain Cleaners 管道疏通剂

Shoe Deodorant 除臭剂

Wax & Polishes 蜡和上光剂

Household Gloves 家用手套

Laundry Products 洗衣产品

Clothes Pegs 衣夹

Hangers & Racks 架子，夹子

Ironing Boards 烫衣板

Laundry Bags & Baskets 洗衣袋，洗衣篮

Laundry Balls & Discs 洗衣球，洗衣片

Scrub Brushes 板刷

Scrub boards 洗衣板

Lint Remover 毛球修剪器

Lint Rollers & Brushes 滚筒式黏毛棒和刷

Mops 拖把

Oversleeves 袖套

Shoe Brush 鞋刷

Sponges & Scouring Pads 擦洗用垫,清洁球和海绵

Squeegees 橡胶清洁刮刀

Waste Bins 垃圾桶

BBQ 烧烤用品

BBQ Accessories 烧烤附件

BBQ Grills 烧烤炉

BBQ Tools 烧烤工具

Bakeware 烘焙用具

Bakeware Sets 烘焙套装

Baking & Pastry Tools 烘焙与面点工具

Baking Dishes & Pans 烘烤盘

Cake Tools 蛋糕工具

Cookie Tools 曲奇工具

Oven Mitts 烤箱用隔热手套

Pie Tools 馅饼工具

Pizza Tools 匹萨工具

Rolling Pins & Pastry Boards 擀面杖和面板

Barware 酒吧酒具

Bar Sets 酒具套装

Bar Tools 酒吧工具

Buckets, Coolers & Holders 冰桶,冷却器和酒架

Cooking Tools 烹调工具

Cheese Tools 乳酪工具

Coffee & Tea Tools 咖啡和茶工具

Colanders & Strainers 滤网

Cooking Tool Sets 烹调工具套装

Dessert Tools 甜品工具

Egg Tools 鸡蛋工具

Fruit & Vegetable Tools 蔬果工具

Herb & Spice Tools 调味工具

Ice Cream Tools 冰激凌工具

Kitchen Lighters 点火器

Kitchen Timers 计时器

Measuring Tools 量具

Meat & Poultry Tools 肉类工具

Mills 研磨器

Openers 开瓶罐器

Pasta Tools 意粉工具

Salad Tools 色拉工具

Seafood Tools 海鲜工具

Specialty Tools 专用工具

Spoon Rests & Pot Clips 汤勺架和锅夹

Sushi Tools 寿司工具

Utensils 器具

Cookware 炊具

Casseroles 砂锅

Cookware Parts 炊具部件

Cookware Sets 炊具套装

Double Boilers 双层蒸锅

Dutch Ovens 荷兰锅

Pans 西式平底锅

Pressure Cookers 压力锅

Soup & Stock Pots 汤锅

Steamers 蒸格

Thermal Cooker 焖烧锅

Water Kettles 烧水瓶和水壶

Woks 中式炒菜锅

Drinkware 饮具

Coffee & Tea Sets 咖啡和茶具套装

Cups & Saucers 茶杯与茶碟

Glass 玻璃杯

Hip Flasks 酒壶

Mugs 马克杯

Tea Accessories 茶具配件

Vacuum Flasks & Thermoses 保温瓶和热水瓶

Water Bottles 水瓶和水壶

Water Pots & Kettles 水罐

Blocks & Roll Bags 刀架和卷袋

Chopping Blocks 砧板

Kitchen Knives 厨房刀

Knife Sets 套装刀具

Sharpeners 磨刀器

Table Decoration & Accessories 桌面装饰品及附件

Mats & Pads 餐具垫

Napkin Rings 餐巾环

Tissue Boxes 纸巾盒

Toothpick Holders 牙签盒

Toothpicks 牙签

Tableware 餐具

Bowls 碗

Chopsticks 筷子

Dinner Knives 餐刀

Dinnerware Sets 餐具套装

Dishes & Plates 餐盘

Flatware Sets 扁平餐具套装

Forks 餐叉

Gravy Boats 调味汁壶

Spoons 调羹

Sugar & Creamer Pots 装糖或乳脂的小罐

Tureens 有盖汤盆

Ashtrays 烟灰缸

Cigarette Box 烟盒

Lighters 打火机

Matches 火柴

Smoking Pipes 烟斗

Aquariums & Accessories 鱼缸及附件

Animal Feeding Supplies 动物饲养用品

Bird Supplies 养鸟用品

Bird Cages 鸟笼

Bird Cups & Bottles 水杯/水瓶

Bird Feeders 喂食用品

Bird Grooming 鸟美容用品

Bird Toy 鸟玩具

Birdcage Covers 鸟笼罩

Training & Behavior Aids 训练用品

Birds 养鸟用品

Cat Supplies 养猫用品

Beds & Mats 猫窝/猫垫

Cat Accessories 猫配饰

Cat Automatic Feeders 猫自动喂食/喂水器

Cat Bedpans 猫便盆

Cat Beds & Mats 猫窝/猫垫

Cat Bowls 猫碗

Cat Cages 猫笼子

Cat Carriers & Bags 猫携带用品

Cat Coats & Jackets 猫外套/夹克

Cat Collars & Leads 猫项圈和牵引

Cat Costumes 角色扮演服

Cat Crates, Gates & Containment 猫大门/围栏

Cat Down & Parkas 猫羽绒服/棉服

Cat Dresses 猫连衣裙

Cat Grooming Supplies 猫美容用品

Cat Hoodies 猫卫衣

Cat Shoes 猫鞋

Cat Toys 猫玩具

Cat Training & Behavior Aids 猫训练用品

Cat Trench 猫风衣

Cat Vests 猫马甲/背心

Furniture & Scratchers 家具及抓爬工具

Dog Supplies 养狗用品

Combs 狗梳子

Dog Accessories 狗配件用品

Dog Automatic Feeders 狗自动喂食/喂水器

Dog Beds/Mats 狗窝/垫子

Dog Caps 狗帽子

Dog Carriers & Bags 狗携带用品

Dog Coats & Jackets 狗大衣/夹克

Dog Collars & Leads 狗项圈和牵引

Dog Down & Parkas 狗羽绒服/棉服

Dog Dresses 狗连衣裙

Dog Feeders 狗喂食/喂水用品

Dog Fences 狗栅栏

Dog Hair Trimmer 电推子

Dog Hoodies 狗卫衣

Dog Jumpsuits & Rompers 狗连体衣/背带裤

Dog Sets 狗套装

Dog Shoes 狗鞋

Dog Shorts 狗短裤

Dog Skirts 狗短裙

Dog Sweaters 狗毛衣

Dog Toys 狗玩具

Dog Trainings 狗训练用品

Dog Trench 狗风衣

Dog Vests 狗背心

Dryers 狗用吹水机

Houses, Kennels 狗舍

Memorials 狗纪念品

Nailclippers 狗指甲钳

Raincoats 狗雨衣

Scissors 毛发剪

Stain & Odor Removers 过滤器

Towels 狗毛巾

Dogs 养狗用品

Farm Animal Supplies 家畜家禽类用品

Cages & Accessories 家禽家畜笼子和附件

Farm Animal Carriers 携带用品

Feeding & Watering Supplies 家禽喂养用品

Farm Animals 家畜家禽类用品

Fish & Aquatic Pet Supplies 鱼和水生生物用品

Air Pumps & Accessories 空气泵及附件

Aquariums 水族箱/鱼缸

FCO2 & Ozone Equipment 二氧化碳和臭氧设备

Cleaning Tools 鱼缸清洁工具

Decorations & Ornaments 装饰品

Filters & Accessories 过滤器及附件

Fish Feeders 喂食/水器

Lightings 照明

Temperature Control Products 控温器材

Water Pumps 养鱼水泵

Fish & Aquatic Pets 鱼和水生生物用品

Horses 马用品

Insect Supplies 昆虫用品

Insects 昆虫用品

Reptiles & Amphibians 爬虫和两栖动物

Wildlife 野生动物

Raincoats 雨衣

Umbrella Stands 伞架

Umbrellas 雨伞

Appliances 家电

Air Conditioners 家用空调

Air Purifiers 空气净化器

Dehumidifiers 干燥器、减湿器

Fans 电风扇

Humidifiers 加湿器

Steam Cleaners 蒸汽清洁器

Ultrasonic Cleaners 超声波清洁器

Vacuum Cleaners 真空吸尘器

Air Conditioner Parts 空调零部件

Air Purifier Parts 空气净化器零部件

Clothes Dryer Parts 干衣机零部件

Dehumidifier Parts 干燥器、减湿器零部件

Electric Heater Parts 电取暖器零部件

Electric Iron Parts 电熨斗零部件

Electric Water Heater Parts 电热水器零部件

Fan Parts 风扇零部件

Freezer Parts 冰柜零部件

Garment Steamer Parts 蒸汽挂烫机零部件

Gas Heater Parts 燃气取暖器零部件

Gas Water Heater Parts 燃气热水器零部件

Hand Dryer Parts 干手机零部件

Heat Pump Water Heater Parts 热泵热水器零部件

Humidifier Parts 加湿器零部件

Ice Maker Parts 制冰机零部件

Kerosene Heater Parts 煤油取暖器零部件

Kitchen Appliance Parts 厨房家电零部件

3 in 1 Breakfast Maker Parts 3 合一早餐机零部件

Blender Parts 搅拌机零部件

Bread Maker Parts 面包机零部件

Chocolate Fountain Parts 巧克力喷泉零部件

Coffee Grinder Parts 咖啡豆研磨机零部件

Coffee Maker Parts 咖啡机零部件

Coffee Roaster Parts 咖啡烘培机零部件

Cooktop Parts 灶台零部件

Crepe Maker Parts 薄煎饼锅零部件

Dish Washer Parts 洗碗机零部件

Disinfecting Cabinet Parts 消毒柜零部件

Egg Boiler Parts 蒸蛋器零部件

Electric Air Pot Parts 电压力暖水瓶零部件

Electric Deep Fryer Parts 电炸锅零部件

Electric Kettle Parts 电水壶零部件

Electric Pressure Cooker Parts 电高压锅零部件

Electric Skillet Parts 电煎锅零部件

Food Mixer Parts 搅拌机零部件

Food Processor Parts 食品加工机零部件

Food Steamer Parts 电蒸锅零部件

Food Waste Disposer Parts 食物垃圾处理器零部件

Hot Plate Parts 电炉零部件

Ice Cream Maker Parts 冰淇淋机零部件

Induction Cooker Parts 电磁炉零部件

Juicer Parts 榨汁机零部件

Meat Grinder Parts 绞肉机零部件

Microwave Oven Parts 微波炉零部件

Oven Parts 烤箱零部件

Pancake Maker Parts 厚煎饼锅零部件

Popcorn Maker Parts 爆米花机零部件

Range Hood Parts 抽油烟机零部件

Range Parts 多功能炉灶零部件

Rice Cooker Parts 电饭煲零部件

Rotisserie Parts 电转烤肉机零部件

Sandwich Maker Parts 三明治机零部件

Slow Cooker Parts 慢炖锅零部件

Toaster Oven Parts 吐司炉零部件

Toaster Parts 吐司机零部件

Vegetable Washer Parts 蔬果清洗机零部件

Waffle Maker Parts 华夫饼机零部件

Yogurt Maker Parts 酸奶机零部件

Refrigerator Parts 冰箱零部件

Solar Water Heater Parts 太阳能热水器零部件
Steam Cleaner Parts 蒸汽清洁器零部件
Steam Press Parts 压力熨烫机零部件
Ultrasonic Cleaner Parts 超声波清洁器零部件
Vacuum Cleaner Parts 真空吸尘器零部件
Washing Machine Parts 洗衣机零部件
Water Dispenser Parts 饮水机零部件
Water Filter Parts 水过滤机/水净化机零部件
Water Softener Parts 水软化机零部件
Wet Towel Dispenser Parts 湿巾机零部件
Wine Refrigerator Parts 冰酒柜零部件
Electric Fireplaces 电壁炉
Electric Heaters 电取暖器
Gas Heaters 燃气取暖器
Hot Water Bottles 热水袋
Kerosene Heaters 煤油取暖器
Blenders 搅拌机
Coffee Grinders 咖啡豆研磨机
Coffee Makers 咖啡机
Coffee Roasters 咖啡烘焙机
Cooking Appliances 烹饪家电
3 in 1 Breakfast Makers 3 合一早餐机
Bread Makers 面包机
Chocolate Fountains 巧克力喷泉
Cooktops 灶
Crepe Makers 薄煎饼锅
Egg Boilers 蒸蛋器
Electric Deep Fryers 电炸锅
Electric Food Steamers 电蒸锅
Electric Grills & Electric Griddles 烧烤机
Electric Pressure Cookers 电高压锅
Electric Skillets 电煎锅
Hot Plates 电炉
Induction Cookers 电磁炉
Microwave Ovens 微波炉
Ovens 烤箱
Pancake Makers 厚煎饼锅
Range Hoods 抽油烟机
Ranges 组合式炉灶

Rice Cookers 电饭煲
Rotisseries 电转烤肉机
Sandwich Makers 三明治机
Slow Cookers 慢炖锅
Toaster Ovens 吐司炉
Toasters 吐司机
Waffle Makers 华夫饼机
Dish Washers 洗碗机
Disinfecting Cabinets 消毒柜
Electric Air Pots 电压力暖水瓶
Electric Kettles 电水壶
Food Mixers 食物搅拌机
Food Processors 食品加工机
Food Waste Disposers 食物垃圾处理器
Ice Cream Makers 冰淇淋机
Ice Crushers & Shavers 碎冰机
Juicers 榨汁机
Meat Grinders 家用绞肉机
Popcorn Makers 爆米花机
Vacuum Food Sealers 家用真空封口机
Vegetable Washers 蔬果清洗机
Yogurt Makers 酸奶机
Clothes Dryers 干衣机
Electric Irons 电熨斗
Garment Steamers 蒸汽挂烫机
Steam Presses 压力熨烫机
Washing Machines 洗衣机
Freezers 家用冷柜
Ice Makers 制冰机
Refrigerators 冰箱
Wine Refrigerators 冰酒柜
Electric Water Heaters 电热水器
Gas Water Heaters 燃气热水器
Heat Pump Water Heaters 热泵热水器
Other Water Heaters 其他热水器
Solar Water Heaters 太阳能热水器
Water Dispensers 饮水机
Water Filters 水过滤机、水净化机
Water Softeners 软水器

Wet Towel Dispensers 湿巾机

Air Cleaning Equipment 空气净化设备

Air Cleaning Equipment Air Shower 风淋室

Industrial Dehumidifier 工业除湿器

Industrial Humidifier 工业加湿器

Air Cleaning Equipment Parts 空气净化器零部件

Air-Compressor Parts 空气压缩机零部件

Air-Compressors 空气压缩机

Animal Husbandry Equipment 饲养管理设备

Animal Feeders 动物饲养设备

Egg Incubators 鸡蛋孵卵器

Apparel Machinery & Parts 服装机械及部件

Button Making Machinery 制扣机

Lace Machines 蕾丝加工机械

Zipper Making Machines 拉链制造机

Apparel Machinery Parts 服装机械部件

Aquaculture Machine Aerators 增氧机

Auxiliary Packaging Machinery 辅助包装设备

Boiler & Parts 锅炉及零部件

Building Material Making Machinery & Parts 建材生产机械及零配件

Brick Making Machinery 制砖机

Cement Making Machinery 水泥加工机

Curb & Paving Stone Forming Machinery 铺路石加工机

Floorboard Making Machinery 地板加工机

Pipe Making Machinery 制管机

Sand Making Machinery 制砂机

Tile Making Machinery 制瓦机

Bus Parts 巴士部件

Bus Body Kits 巴士车身部件

Bus Brakes 巴士刹车部件

Bus Engines 巴士引擎部件

Bus Lights 巴士灯部件

Bus Wheels & Tires 巴士车轮和轮胎

Cap Making Machinery 制帽机械

Chemical Process Equipment Parts 化工过程设备部件

Chemical Product Machinery 化工机械

Chemical Product Machinery Parts 化工机械零部件

Cleaning Equipment 清洁设备

Cleaning Equipment Parts 清洁设备零部件

Construction Machinery & Parts 建筑工程机械及零部件

Asphalt Mixers 沥青混凝土搅拌机

Bulldozers 推土机

Compactors 压土机

Concrete Batching Plant 混凝土搅拌站

Concrete Mixers 混凝土搅拌机

Concrete Pumps 混凝土泵

Concrete Saw 路面锯缝机

Concrete Truck 混凝土搅拌车

Concrete Vibrator 混凝土振捣器

Construction Lifter 施工电梯

Excavators 挖掘机

Loaders 转载机

Motor Graders 平地机

Pavers 铺路机

Pile Driver 打桩机

Power Trowel 抹光机

Rammers 打夯机

Road Roller 压路机

Construction Machinery Parts 建筑机械部件

Consumer Electronics Production Lines 消费电子产品生产线

Energy Saving Equipment 节能设备

Energy Saving Equipment Parts 节能设备零部件

Engine Parts 发动机零部件

Engines 发动机

Evaporator 蒸发器

Farm Machinery & Parts 农用机械及配件

Balers 打捆机

Cultivators 耕作机

Harvesters 收割机

Milking Machines 挤奶机

Oil Pressers 榨油机

Seeders 播种机

Shellers 脱壳机

Tractors 拖拉机

Farm Machinery Parts 农机配件

Feed Processing Machinery 饲料机械

Filter Supplies 过滤用品

Filter Aids 助滤剂

Filter Bags 过滤袋

Filter Cloth 滤布

Filter Films 过滤薄膜

Filter Meshes 过滤网

Filter Papers 滤纸

Food Processing Machinery Parts 食品加工机械零部件

Forestry Machinery 林业器械

Gas Generation Equipment 气体制造机

Gas Generation Equipment Parts 气体制造设备零部件

Glass Processing Machinery 玻璃加工机械

Glove Making Machinery 手套生产机械

Glove Making Machinery Parts 手套生产机械零部件

Home Product Making Machinery 日用品机械

Furniture Making Machinery 家具加工机械

Match Making Machinery 火柴加工机械

Straw Making Machinery 吸管加工机械

Toothpick Making Machinery 牙签加工机械

Wet Wipe Making Machinery 湿纸巾加工机械

Home Product Making Machinery Parts 日杂用品加工机械零部件

Home Textile Product Machinery 家纺生产机械

Felting Machinery 织毯机

Towel Making Machinery 毛巾加工机械

Home Textile Product Machinery Parts 家用纺织品加工机械零部件

Industrial Air Conditioners 工业空调

Industrial Brakes 工业制动器

Industrial Filtration Equipment 工业过滤设备

Air Filter 空气过滤器

Dust Collector 集尘器

Oil Filter 滤油器

Water Filter 水过滤器

Laser Equipment 激光设备

Laser Equipment Parts 激光设备零部件

Leather Production Machinery 皮革制品加工机械

Machine Tools 机床

Bending Machinery 折弯机

Boring Machine 镗床

Drilling Machine 钻床

Gear Cutting Machine 齿轮加工机床

Grinding Machine 磨床

Lathe 车床

Machine Centre 加工中心

Milling Machine 铣床

Planer & Slotting Machine 刨床 / 插床

Punching Machine 冲床

Wire EDM Machine 线切割机床

Accessories 机床附件

Boring Tool 镗刀

Broach 拉刀

Machinery Processing Parts 制造与加工机械零部件

CNC Controller 数控控制器

Chuck 卡盘

Handwheel 手轮

Hob 滚刀

Machine Tool Spindle 机床主轴

Milling Cutter 铣刀

Reamer 铰刀

Tool Holder 刀架

Turning Tool 车刀

Vise 虎钳

Material Handling Equipment 物料搬运设备

Anchor 船锚

Boat Cover 船罩

Boat Engine 船引擎

Bus Accessories 巴士附件

Conveyors 传送带

Cranes 起重机

Forklifts 叉车

Hand Carts & Trolleys 手推车

Hoists 电动葫芦

Lift Tables 升降台

Magnetic Lifter 磁力起重机

Manipulator 机械手

Marine Hardware 船用五金

Marine Propeller 船用螺旋桨

Marine Pump 船用水泵

Pallet Jack 液压车

Stacking Containers 货柜

Stacking Racks & Shelves 货架

Truck Accessories 卡车附件

Wheelbarrows 独轮手推车

Winches 绞盘

Material Handling Equipment Parts 物料搬运设备部件

Metal Processing Machinery 金属加工机械

Metal Casting Machinery 金属铸造机械

Metal Coating Machinery 金属涂镀机械

Metal Cutting Machinery 金属切削机械

Metal Electroplating Machinery 金属电镀机械

Metal Engraving Machinery 金属雕刻机

Metal Processing Machinery Metal Straightening Machinery 金属矫直机

Metallurgy Machinery 冶金机械

Industrial Furnace 工业炉

Rolling Mill 辊轧机

Metallurgy Machinery Parts 冶金机械部件

Mining Machinery 矿山机械

Crusher 破碎机

Dredger 挖泥船

Drilling Rig 钻井机

Mine Mill 磨碎机

Mineral Separator 矿石筛选机

Mining Feeder 矿石进料机

Sand Washer 洗砂机

Vibrating Screen 振动筛

Motor 电机

AC Motor 交流电机

DC Motor 直流电机

Electricity Generation 电力相关设备

Stepper Motor 步进电机

Motor Controller 电机控制器

Motor Driver 电机驱动器

Oil Purifier 油净化器

Packaging Forming Machine 包装成型机械

Packaging Machinery 包装机械

Packaging Machinery Parts 包装机械零部件

Paper Product Making Machinery 纸制品加工机械

Pharmaceutical Machineries 制药机械

Capsule Filling Machine 胶囊填充机

Capsule Polisher 胶囊抛光机

Film Coating Machine 药物包衣机械

Pharmaceutical Packaging Machine 药品包装机械

Tablet Counter 数片机

Tablet Press 压片机

Pharmaceutical Machinery Parts 制药机械零部件

Plastic Processing Machinery 塑料加工机械

Plastic Blowing Machinery 吹塑机

Plastic Crushing Machinery 塑料破碎机

Plastic Cutting Machinery 塑料切割机

Plastic Drying Machinery 塑料干燥机

Plastic Extruders 挤塑机

Plastic Granulators 塑料制粒机

Plastic Injection Machinery 注塑机

Plastic Laminating Machinery 塑料层压机

Plastic Recycling Machinery 塑料回收机

Plastic Rolling Machinery 滚塑机

Plastic Vacuum Forming Machinery 吸塑机

Plastic Product Making Machinery 塑料制品加工机械

Printing Machinery Parts 印刷机械零部件

Post-Press Equipment 印后设备

Pre-Press Equipment 印前设备

Printing Machinery 印刷机械

Refrigeration & Heat Exchange Equipment 制冷及热交换设备

Cold Room 冷冻间

Cooling Tower 冷却塔

Heat Exchanger 热交换器

Ice Machine 制冰机

Industrial Chiller 工业冷凝器

Industrial Freezer 工业冷柜

Industrial Heater 工业加热器

Refrigeration & Heat Exchange Parts 制冷&热交换设备零部件

Rubber Processing Machinery 橡胶加工机械

Rubber Processing Machinery Parts 橡胶加工机械零部件

Rubber Product Making Machinery 橡胶产品加工机械

Shoemaking Machinery 制鞋机械

Shoe Cementing Machines 上胶机(鞋用)

Shoe Eyeleting Machines 打孔机(鞋用)

Shoe Moulding Machines 制模机(鞋用)

Shoemaking Machinery Parts 制鞋机械零部件

Silos 农用储藏仓储

Textile Machinery 纺织机械

Bleaching Machinery 漂白机

Nonwoven Machinery 非织造物机械

Spinning Machinery 纺纱机

Textile Dyeing Machinery 纺织染整机械

Textile Machinery Parts 纺织机械零部件

Truck Parts 卡车部件

Truck Alternator 卡车交流发电机

Truck Battery 卡车蓄电池

Truck Body Parts 卡车车身部件

Truck Brake 卡车刹车部件

Truck Drivetrain & Axles 卡车传动件和车桥件

Truck Engine 卡车引擎部件

Truck Light System 卡车灯件

Truck Starter System 卡车起动系统

Truck Steering 卡车转向部件

Truck Suspension 卡车悬挂部件

Truck Switch 卡车开关

Truck Tire 卡车车胎

Truck Transmission 卡车变速箱部件

Truck Wheel 卡车车轮部件

Ventilation Fan Parts 风机零部件

Ventilation Fans 风机

Welding & Soldering Supplies 焊接配套用品

Electrode Holders 电极夹

Weld Holders 焊钳

Welding Fluxes 焊剂

Welding Helmets 电焊帽

Welding Rods 焊条

Welding Tips 焊接喷嘴

Welding Torches 焊接喷嘴

Welding Wires 焊丝

Welding Equipment 焊接设备

Arc Welders 弧焊机

Butt Welders 对焊机

Electric Soldering Irons 电烙铁

Laser Welders 激光焊机

MIG Welders MIG 焊接机

Plasma Welders 等离子焊机

Plastic Welders 塑料焊机

Press Welders 压焊机

Resistance Welders 电阻焊机

Seam Welders 焊缝机

Spot Welders 点焊机

Tube Welders 管焊接机

Woodworking Machinery 木工机械

Finger Jointer 梳齿机

Press 木工压机

Saw Machinery 锯床

Wood Based Panels Machinery 人造板机械

Wood Boring Machine 木工钻床

Wood Lathe 木工车床

Wood Pellet Mill 木屑颗粒机

Wood Planer 木工刨床

Wood Router 木工雕刻机

Woodworking Benches 木工工作台

Woodworking Machinery Parts 木工机械零部件

Bearing Accessories 轴承附件

Bearing Balls 轴承球

Ball Screws 滚珠丝杠

Bearing Rollers 轴承滚子

Bushings 轴承衬套

Cages 保持架

Housings 轴承座

Bearings 轴承

Ball Bearing 球轴承

Linear Bearings 直线轴承

Needle Bearings 滚针轴承

Pillow Block Bearings 带座轴承

Rod End Bearings 关节轴承

Roller Bearing 滚子轴承

Slewing Bearings 转盘轴承

Sliding Bearings 滑动轴承

Special Bearings 特殊轴承

Custom Fabrication Services 定制加工服务

Casting Services 铸造

Coating Services 涂层

Finishing 表面处理

Forging Services 锻造

Heat Treatment 热处理

Machining Services 机加工

Sheet Metal Fabrication 钣金加工

Welding Services 焊接加工

Cylinders 汽缸/液压缸

Flanges 法兰

Gaskets 垫圈/密封垫

General Mechanical Components 通用机械部件

Hydraulic Parts 液压部件

Linear Guides 直线导轨

Moulds 模具

Pipe Fittings 管接头

Pneumatic Parts 气动元件

Power Transmission Parts 传动件

Gear 齿轮

Gearboxes 齿轮箱

Pulleys 带轮

Shaft Couplings 联轴器

Speed Reducers 减速器

Sprockets 链轮

Transmission Belts 传动带

Transmission Chains 传动链

Universal Joints 万向节

Worms 蜗杆

Pumps 泵

Seals 密封件

Shafts 轴

Valve Parts 阀门配件

Valve Balls 阀球

Valve Bodies 阀体

Valve Caps 阀帽

Valve Needles 阀针

Valve Stems 阀杆

Valves 阀门

Packaging Materials 包装材料

Adhesive Tape, Film, Paper 胶带/不干胶纸/不干胶膜

Adhesives & Sealants　黏合剂及密封剂

Air Dunnage Bag　集装箱充气袋

Aluminum Foil　铝箔

Bottles　包装瓶

Cans　包装罐

Crates　周转箱

Drums, Pails & Barrels　包装桶

Empty Capsules　空胶囊

FIBC Bag　集装袋

Flexi Tank　集装箱液袋

Flower Sleeve　鲜花托

Foil Containers　铝箔容器

Gas Cylinders　气瓶

Gift Ribbon　礼品织带

Handles　把手、提手

Hot Stamping Foil　烫印箔

Jar　广口瓶

Lids, Bottle Caps, Closures　盖子，瓶盖，
　　瓶塞

Mailing Bags　快递信封袋

Packaging Bags　包装袋

Packaging Boxes　包装箱

Packaging Cup, Bowl　包装用杯、碗

Packaging Label　包装标签

Packaging Organza Material　硬珠纱包装
　　面料

Packaging Rope　打包绳

Packaging Tray　包装盘、碟、托

Packaging Tube　管状包装物

Pallets　托盘

Paper & Paperboard　纸、纸张、纸板

Plastic Film　塑料薄膜

Preforms　瓶坯

Protective & Cushioning Material　缓冲、防
　　震防护材料

Strapping　打包带

Printing Materials　印刷材料

Pigment & Dyestuff　染料 & 颜料

Dyestuffs　染料

Paint & Coating　油漆 & 涂料

Pigment　颜料

Printing Mesh　印刷网眼布

Printing Plate　印刷版

Transfer Film　转印膜

Transfer Paper　转印纸

Rubber & Plastics　橡塑原料

Plastic　塑料原材料

Rubber　橡胶原材料

Service Equipment　服务设备

Advertising Equipment　广告设备

Advertising Boards　广告板

Advertising Inflatables　广告气模

Advertising Light Boxes　广告灯箱

Advertising Players　广告播放器

Advertising Screens　广告立牌

Billboards　户外广告牌

Display Racks　广告支架

Poster Materials　布告、海报材料

Roll up Display　易拉宝

Cargo & Storage Equipment　货运仓储设备

Commercial Laundry Equipment　商业清洗
　　设备

Financial Equipment　金融设备

ATM　自动取款机

Bill Counters　点钞机

Coin Counters & Sorters　硬币分类机

Currency Binders　捆钞机

Currency Detectors　验钞机

POS Systems　电子收款机系统

Payment Kiosks　缴费终端

Funeral Supplies　葬礼用品

Restaurant & Hotel Supplies　酒店饭店用品

Cleaning Carts　清洁车

Drink Dispensers　酒水、饮料机

Hotel Amenities　酒店一次性用品

Serving Trays　服务托盘

Store & Supermarket Supplies　超市设备

Checkout Counters　结账柜台

Display Hooks 超市挂钩

Labelers 标签/标签机

Lockers 储物柜

Promotion Table 促销平价台

Refrigeration Equipment 冷冻设备

Shopping Basket 购物篮

Shopping Trolleys & Carts 购物车

Showcase 玻璃柜台

Storage Cages 货框

Supermarket Shelves 超市货架

Trade Show Equipment 展会设备

Banner 展架

Panel Display 屏风展架

Tabletop Display 台式展架

Trade Show Tent 展会帐篷

Truss 架

Vending Machines 自动贩卖机

Jewelry & Watch 珠宝手表

Fashion Jewelry 流行饰品

Anklets 脚链

Bangles 手镯

Beads 珠子

Body Jewelry 身体及穿刺首饰

Bracelets 手链

Brooches 胸针

Charms 小吊坠

Earrings 耳饰

Clip Earrings 耳夹

Dangle Earrings 耳坠

Earring Jackets 耳饰花托

Hoop Earrings 耳圈

Stud Earrings 耳钉

Hair Jewelry 珠宝发饰(带镶嵌,或奢华有首
饰性质的头饰)

Jewelry Sets 首饰套装

Key Chains 钥匙链

Necklace 项链

Pendants 项链吊坠

Rings 戒指

Tie Clips & Cufflinks 领带夹和袖扣

Fine Jewelry 精品珠宝

Anklets 脚链

Beads 珠子

Body Jewelry 身体及穿刺首饰

Bracelets 手链

Bracelets & Cuff 手镯

Bangles 手镯

Brooches 胸针

Charms 小吊坠

Earrings 耳饰

Clip Earrings 耳夹

Drop Earrings 耳坠

Earring Jackets 耳饰花托

Hoop Earrings 耳圈

Stud Earrings 耳钉

Gold Bar 黄金/K 金

Hair Jewelry 珠宝发饰

Jewelry Sets 首饰套装

Key Chains 钥匙链

Loose Gemstones 裸钻

Necklaces 项链

Pendants 项链吊坠

Rings 戒指

Silver Bar 银条

Tie Clips & Cufflinks 领带夹和袖扣

Jewelry Findings & Components 首饰配件
和部件

Jewelry Packaging & Display 首饰包装和展
示用具

Jewelry Tools & Equipment 首饰工具

Other Jewelry 其他首饰

Watches 手表

Pocket & Fob Watches 怀表

Watches Accessories 表附件

Pocket Watch Chains 怀表表链

Repair Tools & Kits 修理工具

Watch Batteries 表电池

Watch Boxes 表盒

Watch Faces　表头

Watch Winders　表络筒机

Watchbands　表带

Wristwatches　腕表

Lights & Lighting　照明灯饰

Commercial Lighting　商业照明

Advertising Lights　广告灯

Laser Flashlight　激光灯

Optic Fiber Lights　光纤灯

Professional Lighting　专业灯具

Stage Lighting Effect　光效灯(舞台灯)

Holiday Lighting　节日照明

Indoor Lighting　室内灯饰灯具

Book Lights　小书灯

Ceiling Fans　吊扇灯

Ceiling Lights　吸顶灯

Chandeliers　枝形吊灯

Desk Lamps　阅读台灯

Downlights　筒灯

Floor Lamps　落地灯

Grille Lights　格栅灯

Night Lights　小夜灯

Pendant Lights　吊灯

Spotlights　射灯(含支架，非灯泡类)

Table Lamps　桌面装饰灯

Track Lighting　轨道灯

Wall Lamps　壁灯

LED Lighting LED 照明

LED Bar Lights LED 硬条灯

LED Modules LED 模块

LED Panel Lights LED 面板灯

LED Spotlights LED 射灯

LED Strip LED 灯带

Led Electronic Candle Led 电子蜡烛

Ballasts　镇流器

Chandelier Crystal　吊灯的水晶

Connectors　连接器

Dimmers　调光器

Flashlight Mount Holders　手电筒固定器

Lamp Bases　灯座

Lamp Covers & Shades　灯罩

Lamp Holder Converters　灯头转换器

Light Beads　灯珠

Lighting Transformers　照明变压器

Lights Lifters　吊灯升降器

RGB Controller　光色控制器

Starters　启辉器、启动器

Switches　开关

Wires & Cables　电线电缆

Lighting Bulbs & Tubes　灯泡、灯管

Energy Saving & Fluorescent　节能、荧光灯泡灯管

Halogen Bulbs　卤素灯泡

High Pressure Sodium Lamps　高压钠灯

Incandescent Bulbs　白炽灯泡

LED Bulbs & Tubes LED 灯泡灯管

Mercury Lamps　水银灯，汞灯

Metal Halide Lamps　金属卤化物灯，金卤灯

Neon Bulbs & Tubes　霓虹灯灯泡灯管，氖灯

Ultraviolet Lamps　紫外线 UV 灯

Xenon Lamps　氙灯

Novelty Lighting　新奇特灯

Outdoor Lighting　室外照明

Floodlights　泛光灯

Landscape Lighting　景观照明

Lantern　灯笼

Lawn Lamps　草坪灯

Lighting Strings　灯串

Path Lights　路径灯

Porch Lights　户外壁灯、廊灯

Solar Lamps　太阳能灯

Street Lights　路灯，街灯

Underground Lamps　埋地灯

Underwater Lights　水下灯具

Portable Lighting　便携式照明

Flashlights & Torches　手电筒

Headlamps　头灯

Portable Lanterns　便携提灯

Portable Spotlights 便携式探照灯

Professional Light 专业灯具

Emergency Lights 应急灯

Grow Lights 植物生长灯

Indicator Lights 指示灯

Industrial Lighting 工业照明

Photographic Lights 业余摄影灯

UV GEL Curing Light 紫外线胶体固化灯

Luggage & Bags 箱包

Backpacks 双肩背包

Bag Parts & Accessories 箱包配件

Briefcases 公文包

Handbags 手提包

Luggage & Travel Bags 旅行箱、旅行包

Luggage 行李箱

Travel Bags 旅行包

Messenger Bags 斜挎包

Special Purpose Bags 特殊用途箱包

Chip Cases 筹码箱

Cosmetic Bags & Cases 化妆包

Hair Scissors Bags 剪刀包

Instrument Bags & Cases 乐器包

School Bags 书包

Shopping Bags 购物袋

Sports & Leisure Bags 运动&休闲包

Cooler Bags 冰袋

Gym Bags 健身包

Lunch Bags 午餐包

Waist Packs 腰包

Wallets & Holders 钱包&卡包

Card & ID Holders 卡包

Coin Purses 零钱包

Key Wallets 钥匙包

Money Clips 钱夹

Wallets 钱包

Mother & Kids 孕婴童

Activity & Gear 婴儿活动用品

Baby Playpens 婴儿围栏

Baby Seats & Sofa 婴儿椅/沙发

Baby Stroller 婴儿推车

Backpacks & Carriers 背婴带

Bouncers, Jumpers & Swings 摇椅/蹦蹦座/秋千

Harnesses & Leashes 学步带

Shopping Cart Covers 购物车座椅

Swimming Pool & Accessories 游泳池和配件

Swimming Pool 游泳池

Walkers 学步车

Baby Care 婴儿护理

Baby Cotton Swabs 婴儿棉签

Bath & Shower Products 洗浴用品

Baby Tubs 婴儿浴盆

Bath Brushes 婴儿浴擦

Shampoo Cap 洗头帽

Towels 婴儿毛巾

Water Thermometers 水温计

Clean Tweezers 清洁镊子

Dental Care 口腔护理

Baby Teethers 婴儿牙胶

Toothbrushes 乳牙刷/训练牙刷

Toothpaste 婴儿牙膏

Ear Syringe 耳朵护理

Grooming & Healthcare Kits 套装

Hair Care 头发护理

Brushes & Combs 梳子

Hair Dryers 婴儿吹风机

Hair Trimmers 婴儿理发器

Shampoo Capes 婴儿理发围布

Nail Care 指甲护理

Nappy Changing 尿布/尿裤用品

Baby Nappies 婴儿尿布/尿裤

Changing Pads & Covers 换尿布垫

Diaper Fixed Belt 尿布带

Nappy Bags 妈咪包

Nappy Liners 尿布衬里

Wet Reminder 尿湿提醒器

Nasal Aspirator 吸鼻器

Potties 便盆/坐便椅

Scales 婴儿体重秤

Skin Care 皮肤护理

Thermometers 体温计

Wet Wipes 婴儿湿巾

Baby Clothing 婴儿服装/配件(0-2 岁)

Bibs & Burp Cloths 围兜围嘴

Gloves & Mittens 手套，连指手套

Hats & Caps 帽子

Headwear 婴儿头饰

Receiving Blankets 婴儿包毯

Baby's Sets 婴儿套装

Blouses & Shirts 衬衫

Bodysuits & One-Pieces 连体衣/爬服

Bodysuits 三角裤爬服

Footies 包脚爬服

Rompers 短裤爬/长裤爬

Dresses 连衣裙

Hoodies & Sweatshirts 卫衣帽衫

Outerwear & Coats Snow Wear 羽绒服/棉

Vests & Waistcoats 马夹 / 外套背心

Pants 长裤

Polo Shirts Polo 衫

Shorts 短裤

Skirts 半身裙

Sleeping Bags 睡袋

Sleepwear & Robes 睡衣，睡袍

Socks & Leg Warmers 袜子/暖腿套

Leg Warmers 暖腿套

Socks 短袜

Tights 连裤袜

Sweaters 毛衣

Swimwear 泳衣

T-Shirts T 恤

Underwear 内衣

Baby Shoes 婴儿鞋

Boots 靴子

Crib Shoes 床鞋

Fashion Sneakers 休闲运动鞋

First Walkers 学步鞋

Leather Shoes 皮鞋

Loafers 乐福鞋-便鞋

Mules & Clogs 花园鞋

Sandals & Clogs 凉鞋花园鞋

Baby Slippers 拖鞋/家居鞋

Sport Shoes 运动鞋

Bedding 婴儿寝具床品

Baby Cribs 婴儿床

Baby Pillows 婴儿枕头

Baby Sleeping Monitors 蹬被提醒器

Bedding Sets 婴儿床上套件

Blanket & Swaddling 婴儿毛毯/裹毯

Bumpers 床围

Cradle 摇篮

Crib Netting 婴儿床网/蚊帐

Pillow Cases 婴儿枕套

Sheets 婴儿床单

Children's Clothing 儿童服装(2 岁以上)

Blouses & Shirts 衬衫

Children's Sets 童装套装

Dresses 连衣裙

Family Matching Outfits 亲子装

Hoodies & Sweatshirts 卫衣帽衫

Jeans 牛仔裤

Outerwear & Coats 外套和大衣

Down & Parkas 羽绒服/棉服

Jackets & Coats 春秋夹克/外套

Trench 风衣

Vests & Waistcoats 马夹/外穿背心

Wool & Blends 毛呢外套

Overalls 背带裤

Sleepwear & Robe 睡衣/睡袍

Blanket Sleepers 连体睡衣/睡袋

Nightgowns 睡裙

Pajama Bottoms 睡裤

Pajama Sets 睡衣睡裤套装

Pajama Tops 睡衣

Robes 睡袍

Socks & Leggings 袜子

Leg Warmers 暖腿套

Leggings 打底裤

Socks 短袜

Tights & Stockings 连裤袜/长筒袜

Suits & Blazers 西服

Sweaters 毛衣

Swim wear 泳衣/沙滩服

Board Shorts 沙滩短裤

Cover-Ups 裹裙/披纱

One Pieces 女孩连体泳衣

Rash Guards 冲浪服/防辐射服

Trunks 男孩泳裤

Two Pieces 女孩分体泳衣

T-Shirts T恤

Underwear 内衣

Long Johns 秋衣秋裤

Panties 内裤

Slips 衬裙

Tanks & Camisoles 吊带/背心

Training Bra 文胸

Undershirts 内穿汗衫

Children's Shoes 童鞋

Athletic Shoes(old) 专业运动鞋

Baseball Shoes 棒球鞋

Basketball Shoes 篮球鞋

Boating Shoes 划船鞋

Bowling Shoes 保龄球鞋

Cycling Shoes 自行车鞋

Dance Shoes 舞鞋

Fitness Shoes 健身鞋

Football Shoes 美式足球鞋

Golf Shoes 高尔夫鞋

Hiking Shoes 登山鞋

Lacrosse Shoes 长曲棍球鞋

Racquetball Shoes 壁球鞋

Rugby Shoes 橄榄球鞋

Running Shoes 跑鞋

Skate Shoes 滑冰鞋，溜冰鞋

Skateboarding Shoes 板鞋

Soccer Shoes 足球鞋

Tennis Shoes 网球鞋

Track & Field Shoes 田径鞋

Volleyball Shoes 排球鞋

Walking Shoes 徒步鞋

Wrestling Shoes 摔跤鞋

Boat Shoes 船鞋

Boots 靴子

Fashion Sneakers 非专业运动鞋

Flats 平跟鞋

Leather Shoes 皮鞋

Loafers 乐福鞋-便鞋

Mules & Clogs 花园鞋

Oxfords 牛津鞋

Sandals 凉鞋

Children Slippers 拖鞋/家居鞋

Feeding 婴儿喂养用品

Baby Food Mills 婴儿食品研磨器

Baby Food Storage 婴儿食品存储盒

Bottle Feeding 奶粉喂养

Bottle Brushes 奶瓶刷

Bottles 奶瓶

Drying Rack 干燥架

Milk Powder Blender 奶粉搅拌器

Nipple 奶嘴

Warmers & Sterilizers 暖奶器和消毒器

Breast Feeding 母乳喂养

Breast Pump Accessories 吸乳器配件

Breast Pumps 吸乳器

Nursing Covers 喂养罩衣

Nursing Pads 防溢乳垫

Insulation Bags 奶瓶保温箱/包

Pacifier 安抚奶嘴

Seats 喂养椅

Booster Seats 加高餐椅

Highchairs 高脚喂养椅

Solid Feeding 食物喂养

Cups 喂养杯/学饮杯

Dishes 婴儿喂养碗

Electric Porridge Pot BB 煲/电粥锅

Solid Feeding Utensils 婴儿喂养匙

Maternity 孕妇装

Blouses & Shirts 雪纺衫/衬衫

Coats 长款外套/大衣/棉服

Dresses 连衣裙

Hoodies 卫衣帽衫

Intimates 贴身衣物

Jackets 夹克/短外套

Jeans 牛仔裤

Leggings 打底裤

Maternity & Nursing Bras 孕妇、哺乳文胸

Pants & Capris 长裤/九分七分五分裤

Shorts 短裤

Skirts 半身裙

Sleep & Lounge 睡衣，家居服

Suits 西装

Sweaters 毛衣

Swimwear 泳衣

Tanks & Camis 背心，吊带

Tees T 恤

Tights & Hosiery 连裤袜，袜子

Vests 外套背心

Safety 婴儿安全防护

Baby Monitors 婴儿监控器

Cabinet Locks & Straps 多功能安全锁

Child Car Safety Seats 婴儿/儿童安全座椅

Edge & Corner Guards 桌角防撞保护

Electrical Safety 防触电保护

Gates & Doorways 安全门/安全护栏

Office & School Supplies 办公文教用品

Books 书籍

Calendars, Planner & Cards 日历，名片，明信片

Business Cards 名片

Calendar 日历

Card Stock 名片册

Greeting Cards 贺卡

Planners 记事本

Correction Supplies 修正用品

Correction Fluid 修正液

Correction Tape 修正带

Eraser 普通橡皮

Office & School Supplies 办公裁剪用品

Cutting Mats 裁纸垫

Letter Opener 开信刀

Scissors 文具剪刀

Utility Knife 美工刀

Desk Accessories & Organizer 桌上收纳用品

Bookends 书立

Card Holder & Note Holder 卡片便签收纳用品

Clip Holder & Clip Dispenser 夹子收纳用品/回形针盒

Desk Set 桌面用具套装

File Tray 桌上文件架

Letter Holder 信件收纳用品

Magazine Organizer 杂志收纳用品

Pen Holders 笔筒

Stationery Holder 文具收纳用品

File Folder Accessories 文件夹附件

Filing Products 文件夹、文件袋等

Labels, Indexes & St amps 徽章，标签，工牌，印章

Badge Holder & Accessories 徽章夹及其附件

Bookmark 书签

Magazines 杂志

Notebooks & Writing Pads 笔记本、拍纸本等书写用品

Clipboard 写字夹板

Memo Pad 便签本

Notebook 笔记本

Padfolio 经理夹

Office Adhesives & Tapes 办公胶水、胶带等

Glue Stick 固体胶棒

Liquid Glue 液体胶水

Office Adhesive Tape 文具胶带

Tape Dispenser 胶带座

Office Binding Supplies 办公装订用品

Binder Index Dividers 索引分页

Binding Combs & Spines 装订耗材

Book Cover 本册产品封面

Clips 夹子

Hole Punch 打孔机

Pin 大头针

Staple Remover 拔钉机

Stapler 订书机

Staples 订书钉

Office Electronic 办公设备，办公电子

All-in-One Printers 一体机

Binding Machine 装订机

Calculator 计算器

Conference System 会议系统

Copiers 复印机

Digital Duplicator 速印机

Electronic Dictionary 电子词典

Fax Machines 传真机

Graph Plotter 绘图机

Laminator 塑封机

Laser Pointers 激光笔

Paper Trimmer 裁纸器

Printer Supplies 打印机耗材和部件

Cartridge Chip 墨盒 / 粉盒芯片

Continuous Ink Supply System 连续供墨系统

Fuser Film Sleeves 定影膜

Fuser Roller 定影辊

nk Cartridges 墨水盒

Ink Refill Kits 填充墨水组

OPC Drum 光导鼓

Printer Parts 打印机部件

Printer Ribbons 打印机色带

Toner Cartridges 墨粉盒、碳粉盒、粉仓

Toner Powder 墨粉、碳粉

Scanners 扫描仪

Shredder 碎纸机

Time Recording 考勤机

Visual Presenter 视屏展示台

Office Furniture 办公家具

Computer Desks 电脑桌

Conference Chairs 会议椅

Conference Tables 会议桌

Filing Cabinets 文件柜

Magazine Racks 杂志架，书报架

Office Chairs 办公椅

Office Desks 办公桌

Office Partitions 办公隔断

Office Sofas 办公室沙发

Other Office Furniture 其他办公家具

Reception Desks 接待台

Other Office & School Supplies 其他办公教
学用品

Painting Supplies 绘画用品

Art Sets 美术成套用具

Calligraphy Brushes 毛笔

Crayons 蜡笔

Easels 画架

Paint Brushes 画笔

Painting Medium 画纸、画布等绘画媒介

Painting Canvas 画布

Painting Paper 画纸

Paints 颜料

Acrylic Paints 丙烯颜料

Oil Paints 油画颜料

Water Color 水彩颜料

Palette 调色板

Paper 办公用纸，纸制品

Carbon Paper 碳式复写纸

Carbonless Paper 无碳复写纸

Cash Register Paper 收银纸

Copy Paper 复印纸

Letter Pad / Paper 信纸

Paper Envelopes 信封

Photo Paper 相片纸

Thermal Fax Paper 热敏传真纸

Pens, Pencils & Writing Supplies 钢笔，铅笔
及书写工具

Ballpoint Pens　圆珠笔

Fountain Pens　钢笔

Gel Pens　中性笔

Markers & Highlighters　马克笔、荧光笔等

Art Markers　美术用马克笔

Highlighters　荧光笔

Marker Pens　马克笔

Whiteboard Marker　白板笔

Multi Function Pen　多功能笔

Pen refill　书写用品附件

Pencil Cases & Bags　笔盒、笔袋等

Pencil Bags　笔袋

Pencil Cases　笔盒

Pencil Sharpeners　削笔用具

Pencils　铅笔

Colored Pencils　彩色铅笔

Mechanical Pencils　自动铅笔

Standard Pencils　普通铅笔

Presentation Boards　书写板、公告板等

Blackboard　黑板

Board Eraser　白板擦、黑板擦等

Blackboard Eraser　黑板擦

Whiteboard Eraser　白板擦

Bulletin Board　公告板

Flip Chart　活动挂图

Whiteboard　白板

School & Educational Supplies　文具和教具

Chalk　粉笔

Drafting Supplies　绘图工具

Compasses　圆规

Math Sets　数学绘图套装

Protractor　量角器

Rulers　尺

Educational Equipment　教学设备

Lab Supplies　实验用品

Beaker　烧杯

Buret　滴定管

Centrifuge Tubes　离心管

Flask　烧瓶

Funnel　漏斗

Lab Balance　实验室天平

Lab Drying Equipment　实验室干燥设备

Laboratory Bottle　实验室瓶

Laboratory Centrifuge　实验室离心机

Laboratory Clamp　实验室夹

Laboratory Cylinder　实验室量筒

Laboratory Heating Equipment　实验室加热设备

Laboratory Refrigeration Equipment　实验室冷冻设备

Laboratory Thermostatic Devices　实验室恒温设备

Petri Dish　陪替式焙养皿

Pipette　吸液管

Test Tube　试管

Map　地图

Stationery Set　文具套装

Stencils　镂花模板

Teaching Resources　教学用具

Chemistry　化学教学用具

Geography　地理教学用具

Language Learning　语言学习教学用具

Mathematics　数学教学用具

Medical Science　医学教学用具

Physics　物理教学用具

Phones & Telecommunications　电话和通信

Communication Equipment　通信设备

Antennas for Communications　通信天线

Communication Cables　通信电缆

Fiber Optic Equipment　光通信设备

Fixed Wireless Terminals　固定无线终端

Telecom Parts　通信附件

Telecommunication Tower　通信塔

Mobile Phone Accessories & Parts　手机配件和零件

Accessory Bundles　配件套装

Armbands　臂带

Mobile Phone Camera Modules　摄像头模块

Mobile Phone Circuits 电路板

Dust Plug 手机防尘塞

External Battery Pack 移动电源

Mobile Phone Holders & Stands 手机支架

Mobile Phone Adapters 手机适配器

Mobile Phone Antenna 手机天线

Mobile Phone Bags & Cases 手机包/手机壳

Mobile Phone Batteries 手机电池

Mobile Phone Cables 手机线材

Mobile Phone Chargers 手机充电器

Mobile Phone Flex Cables 手机排线

Mobile Phone Headphones 手机耳机

Mobile Phone Housings 手机机壳

Mobile Phone Keypads 手机按键

Mobile Phone LCDs 手机显示屏幕

Mobile Phone Lens 手机镜头

Mobile Phone Stickers 装饰贴纸

Mobile Phone Straps 手机挂绳

Mobile Phone Stylus 手机手写笔

Mobile Phone Touch Panel 手机屏幕的触摸面板

Screen Protectors 手机屏保膜

Signal Boosters 中继器信号放大器

sim 卡和配件

Mobile Phone SIM Cards 手机 SIM 卡

Sim Card Readers & Backup SIM 卡读卡器和备份器

Sim Cards Adapterssim 卡适配器

Sim Cards Cutters micro Sim 剪卡器

Mobile Phones 手机

Other Phones & Telecommunications Products 其他电话和通信类产品

PBX 集团电话

Pagers 寻呼机

Telephones 固定电话

Telephones & Accessories 电话和附件

Answering Machines 应答机

Caller ID Boxes 来电显示器

Other Telephone Accessories 其他电话附件

Phone Cards 电话卡

Telephone Cords 电话线

Telephone Headsets 电话听筒、耳机

VoIP Products VoIP 产品

Walkie Talkie 步谈机

Security & Protection 安全防护

Access Control 门禁

Access Control Card 门禁卡

Access Control Card Reader 门禁读卡器

Access Control Keypad 门禁键盘

Access Control System 门禁系统

Audio Door Phone 语音电话

EAS System 商品电子防盗系统

Facial Recognition System 面部识别系统

Fingerprint Access Control 指纹门禁

Video Door Phone 视频电话

Emergency Kits 急救箱

Firefighting Supplies 消防器材

Carbon Monoxide Detectors 一氧化碳探测器

Fire Alarm Control Panel 火警控制板

Fire Blanket 防火毯

Fire Escape Ladders 消防逃生梯

Fire Extinguisher 灭火器

Fire Hose 消防软管

Fire Suit 防火服

Heat Detector 热探测器

Lightning Rod 避雷针

Respirators 消防呼吸器

Smoke Detector 烟雾探测器

Other Security & Protection Products 其他安防产品

Safes 保险柜

Self Defense Supplies 自卫防身用品

Sensors & Alarms 传感器和报警器

Surveillance Products 监控器材

CCTV Accessories 监控器材配件

CCTV Camera Housing 监控器护罩

CCTV Lens 监控镜头

CCTV Monitor 闭路监视器

DVR Card 数字视频采集卡

Surveillance Camera 监控摄像机

Surveillance System 集成监控系统

Surveillance Video Recorder 数字硬盘录像机

Workplace Safety Supplies 劳动保护用品

Chemical Respirator 防化呼吸器

Ear Protector 防护耳罩

Kneepads 护膝

Masks 面具

Safety Clothing 安全防护服装

Safety Cones 安全锥

Safety Gloves 防护手套

Safety Goggles 防护眼罩

Safety Harness 安全带套装

Safety Helmet 安全帽

Warning Tape 安全警示带

Adult Shoes 成人鞋

Boat Shoes 船鞋

Boots 靴子

Fashion Sneakers 非专业运动鞋

Flats 平跟鞋

Loafers 乐福鞋-便鞋

Mules & Clogs 花园鞋

Oxfords 牛津鞋

Pumps 高跟鞋

Sandals & Flip Flops 凉鞋&人字拖

Shoe Accessories 鞋附件

Ice Gripper 防滑冰爪

Inserts 后跟垫/缓冲垫

Insoles 鞋内底/鞋垫

Shoe Care Kit 护鞋套件

Shoe Decorations 鞋配饰

Shoe Horns 鞋拔

Shoe Polish 鞋油

Shoe Trees 鞋撑

Shoelaces 鞋带

Shoes Covers 鞋套

Slippers 室内拖鞋/家居鞋

Additional Pay on Your Order 补邮费/差价

Checkout Link 结账专用链

Coupon 优惠券

Down payment/ Purchasing Agent 定金

Giveaways 赠品

Group Purchase for Furniture 家具团购

Mobile Phone Recharge 手机充值

Overseas Warehouse 海外仓储

Upcoming Products 预售品

Sports & Entertainment 运动及娱乐

Baseball Helmets 棒球头盔

Baseballs & Softballs 棒球及垒球

Chest 护胸

Surface Protection 护面

Bicycle 自行车

Bicycle Accessories 自行车附件

Bicycle Bags & Panniers 自行车包和筐

Bicycle Bell 自行车车铃

Bicycle Bottle Holder 自行车水瓶架

Bicycle Computer 自行车行车电脑

Bicycle Helmet 自行车安全帽

Bicycle Lights 自行车灯

Bicycle Lock 自行车锁

Bicycle Pump 自行车打气筒

Bicycle Racks 自行车托架

Bicycle Stickers 自行车车贴

Bicycle Water Bottle 自行车水壶

Bicycle Parts 自行车部件

Bearings 轴承

Bicycle Brake 自行车刹车

Bicycle Chains & Cleaners 自行车链条 & 洗链器

Bicycle Crank & Chainwheel 自行车曲柄和链轮

Bicycle Derailleur 自行车变速器

Bicycle Fork 自行车前叉

Bicycle Frame 自行车车架

Bicycle Freewheel 自行车飞轮

Bicycle Grips 自行车把套

Bicycle Handlebar 自行车手把

Bicycle Headset 自行车车头碗

Bicycle Hubs 自行车花鼓

Bicycle Pedal 自行车脚踏

Bicycle Saddle 自行车座

Bicycle Seat Post 自行车座管

Bicycle Spoke 自行车车轮辐条

Bicycle Stem 自行车管颈

Bicycle Tires 自行车车胎

Bicycle Wheel 自行车车轮

Bottom Brackets 中轴

Rear Shocks 后减震器

Rims 轮辋

Seat posts Clamps 座管夹

Skewer 轮串

Bicycle Repair Tools 修理工具

Electric Bicycle 电动自行车

Electric Bicycle Part 电动自行车部件

Electric Bicycle Battery 电动自行车电池

Electric Bicycle Motor 电动自行车电机

Self -Balance Unicycle 平衡车

Camping & Hiking 野营及徒步旅行

Camping Knife 野营刀

Camping Mat 野营垫子

Compasses 指南针

Insoles 鞋垫

Outdoor Lighting 户外灯具

Outdoor Stove 户外炉具

Outdoor Tableware 户外餐具

Picnic Bags 野餐包

Professional Climbing Bags 专业登山包

Scarves 围脖头巾

Sleeping Bags 睡袋

Sun Shelter 遮阳棚

Tent Accessories 帐篷配件

Tents 帐篷

Travel Kits 旅行成套工具

Walking Sticks 手杖

Water Bags 水袋

Cheerleading & Souvenirs 啦啦队用品及纪念品

Cheerleading 啦啦队用品

Fan Horns 气喇叭

Pom Poms 绒球，绒花

Whistle 口哨

Souvenirs 纪念品

Entertainment 娱乐

Air Hockey 空气曲棍球

Amusement Park 游乐园设施

Bumper Cars 碰碰车

Climbing Walls 攀登墙

Playground 小型组合游乐园

Slides 滑梯

Water Play Equipment 水上游乐设备

Artificial Grass & Sports Flooring 造草皮及场地地板

Board Game 桌上游戏

Bowling 保龄球用品

Bungee 蹦极

Chess Games 棋类游戏

Coin Operated Games 投币式游艺机

Darts 飞镖用品

Gambling 博彩

GamblingBingo 宾果游戏

GamblingDice 骰子

Gambling Tables 赌博桌

Playing Cards 扑克牌

Poker Chips 扑克筹码

Go Karts 卡丁车

Snooker & Billiard 桌球用品

Balls 台球

Cues 台球球杆

Tables 台球桌

Soccer Tables 桌面足球

Fish Finder 探鱼器

Fishhooks 钓钩

Fishing Bags 钓鱼包

Fishing Chairs 钓鱼椅

Fishing Float 鱼浮

Fishing Lines 钓线

Fishing Lures 鱼饵

Fishing Net 渔网

Fishing Reels 钓轮

Fishing Rods 钓竿

Fishing Rope 渔绳

Fishing Tackle Boxes 钓鱼工具箱

Fishing Tools 钓鱼工具

Fishing Waders 钓鱼防水长靴

Rod Combo 带钓竿的组合套装

Vests 钓鱼背心

Fitness & Body Building 健身及塑形

Boxing 拳击

Boxing Ring 擂台

Punching Bag & Sand Bag 沙包

Punching Balls & Speed Balls 拳击球

Fitness Equipment 健身器材

AB Zones 扭腰椅

Ab Rollers 健腹轮/健腹器

Barbells 杠铃

Dumbbells 哑铃

Fitness Balls 健身球/保健球

Hand Gripper Strengths 臂力器

Hand Grips 握力器

Horizontal Bars 单杠

Horse Riding Machines 骑马机

Hula Hoops 呼啦圈

Indoor Cycling Bikes 动感单车

Integrated Fitness Equipment 综合训练器

Jump Ropes 跳绳

Parallel Bars 双杠

Pedometers 步程计

Power Wrists 腕力器

Push-Ups Stands 俯卧撑架

Racks 器材架

Resistance Bands 拉力带

Row Machines 划船机

Sit Up Benches 仰卧板/健腹板

Steppers 踏步机

Trampolines 蹦床

Treadmills 跑步机

Twist Boards 扭腰盘

Vibration Fitness Massager 甩脂机

Weight Benches 举重床/卧推器

Gymnastics 体操

Martial Arts 武术

Outdoor Fitness Equipment 户外健身设备

Taekwondo & Karate 空手道和跆拳道

Weight Lifting 举重

Yoga 瑜伽

Resistance Bands 瑜伽拉力带

Yoga Bags 瑜伽包

Yoga Balls 瑜伽球

Yoga Belts 伸展带/瑜伽绳

Yoga Blankets 瑜伽铺巾

Yoga Blocks 瑜伽砖

Yoga Circles 瑜伽圈/普拉提圈

Yoga Hair Bands 瑜伽发带

Yoga Mats 瑜伽垫

Club-Making Products 高尔夫球杆部件

Club Grips 高尔夫球杆握把

Club Heads 高尔夫球杆杆头

Club Shafts 高尔夫球杆杆身

Golf Bags 高尔夫球包

Golf Balls 高尔夫球

Golf Carts 高尔夫车

Golf Clubs 高尔夫球杆

Golf Tees 高尔夫球座

Golf Training Aids 高尔夫训练器

Golf Trolley 高尔夫手推车

Horse Racing 赛马用品

Chaps 皮套裤

Halters 马勒、缰绳

Horse Care Products 马匹照顾产品

Horse Rugs 马衣

Horseshoes 马蹄铁

Saddle Pads 马鞍垫

Saddles 马鞍

Hunting 狩猎用品

Blind & Tree Stand 帐篷和树架

Bow & Arrow 弓及箭支

Hunting Bags & Holsters 打猎背包&枪套

Holsters 打猎枪套

Hunting Backpack Bags 打猎背包

Pouches 弹药袋

Waist Bags 打猎腰包

Hunting Cameras 狩猎摄影机

Hunting Decoy 狩猎诱饵

Hunting Gun Accessories 猎枪部件

Hunting Lights 打猎照明设备

Hunting Optics 狩猎用光学器材

Lasers 激光器件

Monocular/Binoculars 单双筒镜

Night Visions 夜视仪

Rangefinders 测距仪

Riflescopes 瞄具

Spotting Scopes 弹着观测镜

Hunting knives 打猎用刀具

Scope Mounts & Accessories 瞄具底座/套环/导轨

Tactical Headsets & Accessories 战术通话设备及配件

Weapon Lights 枪载照明具

Ice Hockey 冰上曲棍球

Ice Hockey & Field Hockey 曲棍球运动

Jerseys 运动衫

Kids' Sneakers 儿童运动鞋

Musical Instruments 乐器

Brass Instruments 铜管乐器

Baritone 低音乐器

Trombone 长号

Trumpet 小号

Keyboard Instruments 键盘乐器

Accordion 手风琴

Electronic Organ 电子琴

Piano 钢琴

Percussion Instruments 打击乐器

Bells & Chimes 铃铛/编钟

Drum 鼓

Gong & Cymbals 锣、钹

Parts & Accessories 零配件

Stringed Instruments 弦乐器

Cello 大提琴及配件

Guitar 吉他

Guzheng 古筝

Stringed Instruments 二胡及配件

Violin 小提琴

Woodwind Instruments 木管乐器

Clarinet 单簧管

Flute 长笛

Harmonica 口琴

Parts & Accessories 零配件

Piccolo 短笛

Saxophone 萨克斯管

Other Sports & Entertainment Products 其他体育娱乐用品

Racquet Sports 球拍运动

Badminton 羽毛球用品

Accessories & Equipment 羽毛球配附件

Badminton Rackets 羽毛球拍

Shuttlecock 羽毛球

Cricket 板球用品

Squash Rackets 壁球拍用品

Table Tennis 乒乓球用品

Table Tennis Balls 乒乓球

Table Tennis Rackets 乒乓球拍

Table Tennis Tables 乒乓球桌

Roller, Skateboard &Scooters 轮滑与滑板运动

Flashing Roller 闪光滚轮

Scooters 滑板车

Scooters Electric Scooters 电动滑板车

Scooters Gas Scooters 气动小摩托车

Handicapped Scooters 残疾人摩托车

Kick Scooters, Foot Scooters 脚踏滑板车

Scooter Parts & Accessories 滑板车部件及附件

Skate Board 滑板

Rugby 橄榄球

Football & Rugby 橄榄球

Shooting 射击用品

Paintball Accessories 彩弹配件

Paintball Markers 彩弹枪

Paintballs 彩弹球

Skiing & Snowboarding 滑雪用品

Skiing 滑雪用品

Ski Poles 滑雪撑杆

Snowboards & Skis 滑雪板

Sleds & Snow Tubes 雪橇及充气轮胎

Sneakers 成人运动鞋

Aqua Shoes 涉水和溯溪鞋

Badminton Shoes 羽毛球鞋

Baseball Shoes 棒球鞋

Basketball Shoes 篮球鞋

Boating Shoes 划船鞋

Bowling Shoes 保龄球鞋

Cycling Shoes 骑行鞋

Dance Shoes 舞鞋

Football Shoes 美式足球鞋

Golf Shoes 高尔夫鞋

Hiking Shoes 登山鞋

Lacrosse Shoes 长曲棍球鞋

Racquetball Shoes 壁球鞋

Rugby Shoes 橄榄球鞋

Running Shoes 跑鞋

Skate Shoes 滑冰鞋，溜冰鞋

Skateboarding Shoes 板鞋

Snowboarding & Skiing Shoes 滑雪鞋

Soccer Shoes 足球鞋

Table Tennis Shoes 乒乓球鞋

Taekwondo Shoes 跆拳道鞋

Tennis Shoes 网球鞋

Toning Shoes 美体塑身鞋

Track & Field Shoes 田径鞋

Volleyball Shoes 排球鞋

Walking Shoes 步行鞋

Water Sports Shoes 水上运动鞋

Wrestling Shoes 摔跤鞋

American Football 美式足球

Baseball 棒球

Bowling 保龄球

Boxing 拳击

Jerseys 运动衫

Robes 长袍

Trunks 平角裤

Camping & Hiking 露营，远足

Camping & Hiking Down 登山/户外羽绒服

Cheerleading Uniforms 啦啦队服

Cycling 自行车

Arm warmers 手臂套

Base Layers 骑行内衣

Bib Shorts 背心短裤

Cycling Sets 骑行服套装

Cycling Shorts 短裤

Face Mask 面罩

Cycling Headwear 骑行头饰

Jackets 夹克

Jerseys 运动衫

Legwarmers 暖脚套

Pants 长裤

Sport Shoe Cover 鞋套

Socks 袜子

Tights & Pants 紧身裤

Vests 背心

Dancing 舞蹈

Golf 高尔夫

Golf Training T Shirts 高尔夫训练体恤

Hunting 打猎

Accessories 配件/配饰

Base Layers 衬里衫

Coats & Jackets 外套夹克

Ghillie Suits 隐秘猎装

Motorcycle & Auto Racing 摩托车/汽车赛车

Running 赛跑

Arm Warmers 防晒臂套

Running Bags 跑步专用包

Running Sets 跑步套装

Running Vests 背心

Running T-Shirts T恤

Tights 连裤袜

Skateboarding 滑板

Hoodies 帽衫

Skiing 滑雪

Bibs 围兜

Snowboarding 单板滑雪

Soccer Sets 足球服套装

Training Pants 训练裤

Sports Bras 运动内衣

Sports Caps 运动帽

Baseball Sport Caps 专业棒球帽

Beach Caps 沙滩运动帽

Cycling Caps 骑行帽

Fishing Caps 钓鱼帽

Golf Caps 高尔夫帽

Golf Visors 鸭舌帽

Hiking Caps 登山帽

Hunting Caps 打猎帽

Running Caps 跑步帽

Swimming Caps 泳帽

Tennis Caps 网球帽

Sports Eyewear 运动眼镜

Cycling Eyewear 骑行眼镜

Fishing Eyewear 钓鱼眼镜

Hiking Eyewear 登山眼镜

Skiing Eyewear 滑雪眼镜

Swim Eyewear 游泳眼镜

Sports Gloves 运动手套

Baseball Gloves 棒球手套

Boxing Gloves 拳击手套

Cycling Gloves 骑行手套

Fishing Gloves 钓鱼手套

Goalie Gloves 运动员手套

Golf Gloves 高尔夫手套

Hiking Gloves 登山手套

Hunting Gloves 打猎手套

Racing Gloves 赛车手套

Riding Gloves 马术手套

Running Gloves 跑步手套

Skiing Gloves 滑雪手套

Swimming & Diving Gloves 游泳及潜水手套

Sports Safety 运动护具

Ankle Support 护踝

Arm Warmers 手臂套

Back Support 护背

Elbow & Knee Pads 护肘，护膝

Helmets 头盔

Shin Guard 护胫

Sweatband 防汗带

Waist Support 护腰

Wrist Support 护腕

Swim 游泳

Body Suits 紧身衣

Briefs 短裤

Jammers 防护裤

One-Piece Suits 体式

Rash Guards 冲浪服

Swimming Bags 游泳收纳包

Swimming Towels 快干游泳毛巾

Two-Piece Suits 分体式

Wetsuits 潜水服

Table Tennis 乒乓球服

Jerseys 运动衫

Sets 套装

Shorts 短裤

Tennis 网球

Dresses 连衣裙

Tennis T Shirts 网球体恤

Volleyball 排球

Water Sports 水上运动

Nose/Ear Clips 鼻夹/耳塞

Surfing & Beach Shorts 冲浪&沙滩短裤

Surfing & Beach T-Shirts 冲浪&沙滩 T 恤

Yoga 瑜伽

Women's Yoga sets 瑜伽套装

Tennis 网球用品

Accessories 配件

Tennis Ball 网球

Tennis Rackets 网球拍

Water Sports 水上运动

Beach Frisbees 沙滩飞盘

Beach Kites 沙滩风筝

Racing Boats 赛艇

Rowing Boats 划艇

Surfing 冲浪运动

Swimming & Diving 游泳及潜水

Air Mattresses 空气垫

Diving Masks 潜水面具

Pool & Accessories 泳池及附件

Snorkels 水下通气管

Swimming Fins 蛙鞋

Swimming Rings 泳圈

Water Safety Products 水上救生产品

Life Buoy 救生圈

Life Raft 救生筏

Life Vest 救生衣

Abrasive Tools 磨具

Abrasives 磨料

Construction Tools 建筑工具

Caulking Gun 压胶枪

Glass Cutter 玻璃刀

Ladder & Scaffolding Parts 施工设施部件

Ladders 梯子

Plaster Trowel 抹泥板

Putty Knife 油灰刀

Hand Tools 手工具

Axe 斧

Brush 刷

Chisel 凿

Files 锉

Hammer 锤

Knife 刀

Locksmith Supplies 锁匠工具

Pliers 钳

Saw 锯

Scissors 剪

Screwdriver 螺丝刀

Tap & Die 丝锥板牙

Wrench 扳手

Lifting Tools 起重工具

Jacks 千斤顶

Lifting Sling 吊索

Measurement & Analysis Instruments 仪器仪表

Analyzers 分析仪器

Carbon Analyzers 碳分析仪

Concentration Meters 浓度计

Densitometers 密度计

Gas Analyzers 气体分析仪

Moisture Meters 湿度计

PH Meters PH 计

Counters 计数器

Electrical Instruments 电工仪器仪表

Battery Testers 电池测试器

Clamp Meters 钳型表

Current Meters 电流测量仪表

Energy Meters 电能仪表

Frequency Meters 频率测量仪表

Multimeters 万用表

Potentiometers 电位差计

Power Meters 功率测量仪表

Resistance Meters 电阻测量仪表

Voltage Meters 电压测量仪表

Electronic Measuring Instruments 电子测量仪器

Industrial Metal Detectors 金属探测仪

Oscilloscopes 示波器

Signal Generators 信号发生器

Spectrum Analyzers 频谱分析仪

Flow Measuring Instruments 流量测量仪器

Flow Meters 流量计

Flow Sensors 流量传感器

Gas Meters 煤气表

Water Meters 水表

Instrument Parts & Accessories 配件与附件

Level Measuring Instruments 物位测量仪器

Instrument Design Services 仪器仪表设计服务

Instrument Stocks 仪器仪表库存

Measuring & Gauging Tool 量具

Dial Indicators 百分表/千分表

Gauge 量规

Micrometers 千分尺

Tape Measures 卷尺

Vernier Calipers 游标卡尺

Optical Instruments 光学仪器

Optical Instruments Laser Levels 激光水平仪

Optical Instruments Laser Rangefinders 激光测距仪

Optical Instruments Lenses 透镜

Optical Instruments Magnifiers 放大镜

Optical Instruments Microscopes 显微镜

Optical Instruments Optical Filters 滤光镜，滤光片

Optical Instruments Prisms 棱镜

Optical Instruments Refractometers 折射仪

Optical Instruments Spectrometers 光谱仪

Optical Instruments Telescope & Binoculars 望远镜

Optical Instruments Theodolites 经纬仪

Other Measuring & Analyzing Instruments 其他仪器仪表

Physical Measuring Instruments 物理量测量仪器

Force Measuring Instruments 测力仪器

Hardness Testers 硬度计

Height Measuring Instruments 高度测量仪器

Speed Measuring Instruments 速度测量仪器

Width Measuring Instruments 厚度测量仪器

Pressure Measuring Instruments 压力测量仪器

Pressure Gauges 压力表/压力计

Pressure Monitors 压力监控器

Pressure Regulators 减压器

Pressure Sensors 压力传感器

Pressure Measuring Instruments Pressure Transmitters 压力变送器

Temperature Instruments 温度测量仪器

Testing Equipment 试验机

Timers 计时器

Weighing Scales 衡器，秤

Other Tools 其他工具

Power Tool Accessories 动力工具配件

Power Tools 动力工具

Blower 鼓风机

Drill Bit 钻头

Electric Drill 电钻

Electric Hammer 电锤

Electric Pipe Threader 电动套丝机

Electric Planer 电刨

Electric Saw 电锯

Electric Screwdriver 电动螺丝刀

Electric Trimmer 修边机

Electric Wrench 电动扳手

Glue Gun 胶枪

Grinder 研磨工具

Grinding Wheel 砂轮片

Heat Gun 热风枪

Hot Melt Glue Sticks 热熔胶棒

Hydraulic Tools 液压工具

Nail Gun 钉枪

Pneumatic Tools 气动工具

Polisher 抛光机

Polishing Pad 抛光垫

Sander 砂光机

Sanding Disc 砂光盘

Saw Blade 锯片

Spray Gun 喷枪

Tool Parts 工具零件

Tool Sets　组合工具

Hand Tool Sets　手动工具组合

Paint Tool Sets　油漆工具组合

Power Tool Sets　动力工具组合

Tools Packaging　工具收纳整理

Tool Bag　工具包

Tool Box　工具盒

Tool Cabinet　工具柜

Tool Case　工具箱

Toys & Hobbies　玩具

Action & Toy Figures　手办/可动人形/机器人

Baby Toys　婴儿玩具

Baby Rattles & Mobiles　婴儿摇铃/床铃

Play Mats　婴儿游戏垫

Classic Toys　经典玩具

Balloons　气球

Bath Toy　沐浴玩具

Kaleidoscope　万花筒

Magic Tricks　魔术玩具

Noise Maker　敲击发声玩具

Pretend Play　过家家玩具

Doctor Toys　仿真医药箱/医生玩具

Furniture Toys　仿真家具玩具

Kitchen Toys　仿真厨房类玩具

Tool Toys　仿真维修工具玩具

Spinning Top　陀螺

Sticker　贴纸贴画

Wind Up Toys　发条玩具

Windmill　风车

Yoyo　溜溜球

Diecasts & Toy Vehicles　非遥控类交通工具
　　玩具

Dolls & Stuffed Toys　娃娃/填充及毛绒玩具

Dolls & Accessories　人形娃娃及配件

Doll Houses　娃娃房子

Dolls　人形娃娃

Dolls Accessories　娃娃配件

Stuffed Animals & Plush　填充动物/毛绒玩具

Movies & TV　电影动漫周边填充/毛绒玩具

Plush Backpacks　毛绒背包

Puppets　手偶

Stuffed Animals & Plush Stuffed & Plush
　　Animals　填充/毛绒动物

Stuffed Animals & Plush Stuffed & Plush
　　Plants　填充/毛绒植物

Electronic Toys　电子玩具

Electronic Pets　电子宠物/非遥控类机器宠物

Toy Cameras　玩具相机

Toy Phones　玩具电话/手机

Toy Walkie Talkies　玩具对讲机

Learning & Education　益智玩具

Drawing Toys　画图玩具

Learning Machine　学习机

Math Toys　算术玩具

Playdough　橡皮泥

Readings　读物

Card books　卡片书

Cloth Books　布书

Toy Musical Instrument　玩具乐器

Models & Building Toy　模型，积木和拼插
　　玩具

Blocks　积木

Model Building Kits　模型拼装套件

Novelty & Gag Toys　新奇搞怪玩具

Finger Skateboards & Bikes　手指滑板和自
　　行车

Gags & Practical Jokes　搞怪玩具

Glow in the Dark Toys　夜光玩具

Light-Up Toys　发光玩具

Solar Toys　太阳能玩具

Other Toys & Hobbies　其它玩具

Outdoor Fun & Sports　户外玩具

Bubbles　吹泡泡

Fishing Toys　钓鱼玩具

Flying Disk & Arrow　飞盘和飞箭

Inflatable Toys　充气玩具

Kites & Accessories　风筝/风筝配件

Ride On Toys　可骑乘玩具

Ride On Animal Toys 可骑乘动物玩具

Ride On Cars 可骑乘玩具车

Toy Balls 玩具球

Toy Guns 玩具枪

Toy Sports 运动类玩具

Toy Swings 玩具秋千

Toy Swords 玩具刀剑

Toy Tents 玩具帐篷

Puzzles & Magic Cubes 拼图/立体拼图/七巧板/数独/魔方

Magic Cubes 魔方

Puzzles 拼图/立体拼图/七巧板/数独

Remote Control 遥控玩具

RC Airplanes 遥控飞机

RC Animals 动物遥控

RC Boats 遥控船

RC Cars 遥控车

RC Helicopters 遥控直升机

RC Motorcycles 遥控摩托车

RC Submarine 遥控潜水艇

RC Tank 遥控坦克

RC Trains 遥控火车

RC Trucks 遥控卡车

Simulators 遥控飞行模拟器

Travel and Vacations 旅游及度假

Ctrip 携程

Coupons 优惠券

Ctrip Hotels 携程酒店

Public Transport Discount Cards 交通卡券

Rentals 预定

Book a Rental Bike 自行车

Book a Rental Car 出租车

Hotels 酒店/宾馆

Performances 演出门票

Special Attractions 景点门票

Tour Guides 导游

Travel Discount Coupons 旅游优惠卡券

Annual Discount Cards 旅游年票/一卡通/年卡/套票

Hong Kong & Macau Tourism 港澳送关服务

Hotel Discount Coupons 酒店/宾馆优惠券

Meal Discounts 餐饮优惠券

Shopping Discount Cards 购物折扣卡券

Special Attractions Coupons 景点门票

Tour Guide Services 导游服务

Travel Agency Discount Coupons 旅行社优惠券

Travel Products 旅游相关商品

Flights 机票增值产品

Maps 地图

Other 其他

Postcard 明信片

Souvenirs 旅游纪念品

Traditional Ritual Services 代客烧香/还愿

Travel Essentials 旅游必备品

Travel Routes and Tips 旅游攻略

Appendix 2　服装类专业用语

一、Clothing Terms 服装专业术语

accordion pocket　风琴袋

armhole　袖孔

band　压条

batwing sleeve　蝙蝠袖

bellows pocket　吊袋

bias strip　滚边

binding　滚条

blackstripe　黑条纺

boat neckline; slit neckline; off neckline　一字领口

box pleated pocket　明裥袋

breast pocket　手巾袋

burnt-out　烂花

bust dart　前肩省

bust　胸围

button loop　滚眼

button placement　扣位

button position　扣位，纽扣的位置

button-hole space　眼距

buttonhole spacing　眼距

buttonhole　扣眼

cambric　帆布

card pocket　卡袋

check　格子

chiffon　雪纺

closure　(锁眼的衣片)门襟

collar edge　领外口

collar stand, collar band　底领

collar tab　领袢

collar　衣领

convertible collar　两用领也叫开关领

cotton　棉

cuff　袖头，指装袖头的小袖口

cutting drawing　服装裁剪图

design drawing　款式设计图

double breasted　双排扣

double cuff; French cuff; tumup-cuff; fold-back cuff　双袖头

double welt pocket　双嵌线袋

double-layer　双层

effect drawing　服装效果图

effect　外观，声响，印象，效果

elastic cuff　橡皮筋袖口

emptystripe　空齿纺

fabric　面料

facing　挂面

faille　花瑶

filament　长丝

flange　耳朵皮

flap　袋盖

flare sleeve; trumpet sleeve　喇叭袖

fold line for lapel　驳口

fold line of collar　领上口

french tack　线袢

front cut　止口圆角

front edge　门襟止口

front edge　止口

front fly; top　门襟

front interlining　前身衬

front waist dart　前腰省

front yoke　前过肩

georgette　乔其纱

glasses pocket　眼镜袋

gorge line　串口

hanger loop　吊袢

hem　下摆，指衣服下部的边沿部

imitated silk fabric　仿真丝

insert pocket　插袋

inside pocket　里袋

interlining　衬布

inverted pleated pocket　暗裥袋

jacquard　提花

jeanet　牛仔布

koshibo　高丝宝

lantern sleeve; puff sleeve　灯笼袖

lapel　翻领

lining　里料

lotus leaf collar　荷叶边领

lustrine　绡

mandarin collar　中式领

mock button hole　假眼

modeling　服装造型

neck dart　领省，指在领窝部位所开的省道

sweetheart neckline; heart shaped neckline　鸡心领口

notch lapel　平驳头

notch　领嘴

nylon/polya mide　锦纶

organdy　玻璃纱

overlap　搭门

oxford　牛津布

P/C　涤棉

patch pocket with flap　有盖贴袋

patch pocket　贴袋

peachmoss　绉绒

peachskin　水洗绒/桃皮绒

peacht will　卡丹绒

peak lapel　戗驳头

petal sleeve　花瓣袖

placket　门襟翻边

pocket dart　胁省

pointed collar; peaked collar　尖领

polyester fiber　聚酯纤维

polyester　涤纶

pongee　春亚纺

puff sleeve　泡泡袖

raglan sleeve　连肩袖

raglan sleeve　连袖

rayon　粘胶纤维

rib-knit cuff　罗纹袖口

round collar　圆角领

round neckline　圆领口

satin/charmeuse　缎面

set-in sleeve　圆袖

shawl collar　青果领

shirt collar　衬衫领或衬衣领

shirt sleeve　衬衫袖

shoulder seam　肩缝

shoulder tab; epaulet　肩袢

side dart　横省

side seam　摆缝

silhouette　服装轮廓

silk　真丝

single breasted　单排扣

single welt pocket　单嵌线袋

sleeve opening　袖口

sleeve placket　袖衩条

sleeve slit　袖开衩

sleeve tab　袖袢

sleeve　袖子

spandex/elastic/strec/lycra　氨纶

split raglan sleeve　前圆后连袖

square collar　方领

square neckline　方领口

stand collar; Mao collar　立领

stripe　条子

structure line　服装结构线

structure　结构；构造

style　样式；款式

swallow collar；wing collar　燕子领

T/R　涤捻

Taffeta　塔夫绸

taslon　塔丝隆

top collar stand　领里口

top sleeve　大袖，外袖

tuck　太可褶

twill　斜纹

two-tone　双色

under fly　里襟

under line of collar　领下口

under sleeve　小袖，内袖或里袖

underfly　（钉扣的衣片）里襟

viscose　人造棉

waist tab　腰袢

waist　（衣服的）腰部，腰

waistbelt; waistband　腰带

whitestripe　白条纺

zhongshan coat collar　中山服领

zigzag inside pocket　锯齿形里袋

二、出口服装常用表达方式

		服装统称	garment, clothing, apparel, dress, wear
服装		男装	men's wear, men's suits
		女装	women's wear, women's suits
		童装	children's wear
上装	礼服	晚礼服	evening dress, evening wear
		裙装	dress
	休闲服	休闲装	casual wear, leisure wear
		运动装	sports wear
	外衣	外套	outwear, over coat, jacket
		大衣	coat, overcoat, top coat, casual coat, dressy coat, parka
		风衣、雨衣、风雨衣	coat, raincoat, all weather coat
		套装	suits, sets
	衬衫	硬领衬衫	shirt
		软领花式女衬衫	blouse

续表

下装	裤子	裤子统称	trousers, pants
		短裤	shorts
		宽松裤	slacks
	裙子	裙子统称	skirt
		A 字裙	A-line skirt
		褶裥裙	pleated skirt
		多片裙	gored skirt
		直筒裙	straight skirt
		包裹裙	wrap skirt
内衣		内衣统称	under wear, intimate apparel, foundation garment
		内裤	panties
		背心	vest, halter
		三角裤	briefs
		吊带衬裙	slip
		衬裙	waist slip, half slip
		文胸	bras, corset
针织服装		套头毛衣	sweater, pullover, jumper
		开襟毛衣	cardigan
		T 恤衫(针织翻领衫)	knitted shirt
		针织圆领衫	T-shirt

三、鞋子尺码对照表

美国码 USA	4.5	5	5.5	6	6.5	7	7.5	8	8.5	9	9.5	10	10.5	11
英国码 UK	4	4.5	5	5.5	6	6.5	7	7.5	8	8.5	9	9.5	10	10.5
法国码 EUR	35	36	37	38	39	40	41	42	43	44	45	46	47	48
中国 CHN	225	230	235	240	245	250	255	260	265	270	275	280	285	290

四、常用长度换算表

米制	1 公里 = 1 千米(km) = 1000 米(m) 1 米(m) = 10 分米(dm) 1 分米(dm) = 10 厘米(cm)
英制	1 英尺(ft) = 12 英寸(in) 1 英里(mile) = 5280 英尺(ft) 1 海里(n mile) = 1.1516 英里(mile) 1 码 = 3 英尺(ft) 1 英里(mile) = 1760 码(yards)
米制与英制换算	1 千米(km) = 0.621 英里(mile) 1 英里(mile) = 1.609 千米(km) 1 米(m) = 3.281 英尺(ft) = 1.094 码(yd) 1 厘米(cm) = 0.394 英寸(in) 1 英寸(in) = 2.54 厘米(cm) 1 海里(n mile) = 1.852 千米(km)

缩略词：

yd = 码 yards   ft = 英尺 feet   in = 英寸 inches   km = 千米 kilometers

m = 米 meters   cm = 厘米 centimeters   dm = 分米 decimeters

五、上衣常用尺码表

SIZE INFORMATION 尺码信息				
Size 尺码	Bust 胸围	Adjustable Range 可调范围	Waist 腰围	Adjustable Range 可调范围
S	80CM	77CM~83CM	63CM	60CM~66CM
M	84CM	81CM~87CM	67CM	64CM-70CM
L	87CM	84CM~90CM	70CM	67CM-73CM
XL	90CM	87CM~93CM	74CM	71CM-77CM
2XL	93CM	90CM~96CM	77CM	74CM-80CM
3XL	96CM	93CM~99CM	80CM	77CM-83CM
4XL	102CM	99CM~105CM	90CM	87CM-93CM
5XL	108CM	105CM~111CM	96CM	92CM-100CM
6XL	116CM	113CM-119CM	104CM	100CM-108CM
7XL	124CM	121CM~127CM	112CM	108CM-116CM
8XL	132CM	129CM~135CM	120CM	116CM-124CM

六、下装常用尺码表

SIZE: POINTS OF MEASUREMENTS	S	M	L	1X	2X	3X	4X
PANTS LENGTH BELOW WB	38 1/4	38 3/4	39 1/4	39 3/4	40 1/4	40 3/4	41 1/4
INSEAM	28 1/8	28 1/8	28 1/8	28 1/8	28 1/8	28 1/8	28 1/8
WAIST RELAX	28 3/4	30 1/4	32	33 1/2	35	36 1/2	38
FRONT WAISTBAND SS TO SS (RELAXED)							
WAIST EXTENDED				42			
LOW HIP 8" BB (3 PTS)	40 1/2	42	43 1/2	45 1/4	47	48 3/4	50 1/2
FRONT RISE INCLUDING WB	13	13 1/2	14	14 1/2	15	15 1/2	16
BACK RISE INCLUDING WB	16	16 1/2	17	17 1/2	18	18 1/2	19
THIGH 1" BELOW CROTCH	25	26	27	28	29	30	31
KNEE 11" BELOW CROTCH	20	21	22	23	24	25	26
LEG OPENING	14 1/2	15 1/2	16 1/2	17 1/2	18 1/2	19 1/2	20 1/2

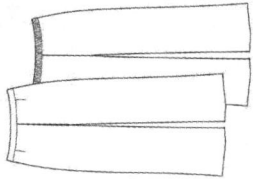

参 考 文 献

[1]　吴思乐，胡秋华. 世纪商务英语谈判口语[M]. 大连:大连理工大学出版社，2015.

[2]　姜继红，崔立标，胡雁峰.Wish 平台操作实训教程[M]. 西安:西安电子科技大学出版社，2020.

[3]　韩乃臣. 外贸企业商务谈判实训手册[M]. 北京：中国人民大学出版社，2014.

[4]　https://language. chinadaily. com. cn/trans/2014-07/25/content_17926524. htm.

[5]　https://www. douban. com/group/topic/29687486/.

[6]　HUGHES J，COOK R，PEDRETTI M，et al. 新编剑桥商务英语(初级)学生用书[M]. 3 版. 北京：经济科学出版社，2018.

[7]　张翠萍. 商贸英语口语一本通[M]. 北京：对外经济贸易大学出版社，2000.

[8]　姜宏. 国际贸易实务与综合模拟实训[M]. 北京：清华大学出版社，2008.

[9]　耿民. 商务英语谈判模拟实训教程[M]. 北京：北京出版社，2008.

[10]　张立玉. 商务谈判英语[M]. 武汉：武汉出版社，2009.

[11]　凌双英，王俊. 实用经贸英语口语[M]. 北京：高等教育出版社，2006.

[12]　张天桥. 国际贸易单证实务与实训[M]. 北京：北京师范大学出版社，2009.

[13]　秦川. 外贸英语会话[M]. 北京：中国对外经济贸易出版社，2006.

[14]　朱文忠，刘平. 新编国际经贸英语[M]. 北京：对外经济贸易大学出版社，2008.

[15]　徐宪光. 商务沟通[M]. 北京：外语教学与研究出版社，2001.

[16]　廖瑛，莫再树. 国际商务英语语言与翻译研究[M]. 2 版. 北京：对外经济贸易大学出版社，2007.

[17]　朱文忠，周杏英. 实用商务谈判英语[M]. 北京：对外经济贸易大学出版社，2007.

[18]　窦卫霖. 跨文化商务交流案例分析[M]. 北京：对外经济贸易大学出版社，2007.

[19]　诸葛霖，江春. 外贸英语对话[M]. 3 版. 北京：对外经济贸易大学出版社，2007.

[20]　廖国强. 经贸英语[M]. 北京：高等教育出版社，2005.

[21]　王珍. 出口贸易操作[M]. 杭州：浙江大学出版社，2007.

[22]　WTO 研究编写组.WTO 国际规则、惯用现用现查[M]. 呼和浩特:内蒙古人民出版社，2002.

[23]　冯祥春. 国际商务英语大词典[M]. 北京：中国对外经济贸易出版社，2000.

[24]　戚建平，邬江兴. 英汉/汉英商务缩略语大词典[M]. 北京：国防工业出版社，2009.

[25]　盛湘君. 跨境电商交际英语[M]. 北京：外语教学与研究出版社，2016.

[26]　Robert M. March. 跨文化商务交流[M]. 北京：对外经济贸易大学出版社，2008.

[27]　刘筱琳. 服装职业英语[M]. 上海：东华大学出版社，2010.

[28]　陈雁，李莉. 服装进出口贸易[M]. 上海：东华大学出版社，2008.

[29]　包振华，张耘. 纺织品外贸单证实务[M]. 北京：化学工业出版社，2014.

[30]　周叔安. 汉英英汉服装分类词汇[M]. 3 版. 北京：中国纺织出版社，2018.